High Power Laser Handbook

Kruger Brentt
Publishers

High Power Laser Handbook

Caner Deger
Federico Suarez

Kruger Brentt
Publishers

2025

Kruger Brentt Publishers UK. LTD.
Company Number 9728962

Regd. Office: 68 St Margarets Road, Edgware, Middlesex HA8 9UU

ISBN: 978-1-78715-381-3

For information on all our publications visit our website at http://krugerbrentt.com/

PREFACE

High power lasers represent one of the most powerful and versatile tools in modern science and engineering, with applications ranging from fundamental research to industrial manufacturing, defense systems, and medical treatments. With their intense beams of coherent light, high power lasers enable precision machining, materials processing, laser spectroscopy, and numerous other cutting-edge applications. Understanding the principles, technologies, and applications of high power lasers is essential for researchers, engineers, and practitioners seeking to harness the full potential of laser technology for scientific discovery, technological innovation, and industrial advancement.

This handbook, titled "**High Power Laser Handbook**" is a comprehensive exploration of the principles, technologies, and applications of high power lasers. Drawing upon the expertise of laser physicists, engineers, and industry experts, this volume offers theoretical insights, practical guidance, and case studies for understanding and utilizing high power lasers in various domains.

From the fundamentals of laser physics and beam generation to the latest advancements in laser amplification, beam shaping, and nonlinear optics, the "High Power Laser Handbook" covers a wide range of topics essential for mastering the field of high power laser technology. Through clear explanations, mathematical derivations, and practical examples, readers will gain a deep understanding of the principles and techniques used to generate, manipulate, and control high power laser beams.

Moreover, this handbook explores the applications of high power lasers across different disciplines and industries, including laser materials processing, laser-based manufacturing, laser spectroscopy, medical lasers, and defense systems. By examining case studies, research findings, and industrial applications, we highlight the diverse ways in which high power lasers are used to solve real-world problems, drive technological innovation, and create economic value.

As we continue to push the boundaries of laser technology and explore new frontiers in science and engineering, the importance of high power lasers in driving innovation and advancing knowledge has never been greater. It is our hope that the "High Power Laser Handbook" will serve as a valuable resource for researchers, engineers, and students in the field of laser science and technology. Whether you are designing laser systems, conducting research in laser physics, or applying lasers in industrial settings, this handbook invites you to explore the exciting world of high power lasers and unlock new possibilities for scientific discovery and technological advancement.

Caner Deger
Federico Suarez

CONTENTS

CHAPTER-1

LASER SYSTEMS OF HIGH POWER IN THE UV AND VISIBLE SPECTRAL RANGES

1. INTRODUCTION: AN OVERVIEW

Laser systems utilizing noble-gas halides are the most potent and efficient generators of coherent radiation within the ultraviolet (UV) spectrum. Currently, these systems function as an exclusive method for resolving essential and practical issues, such as inertial nuclear fusion, the physics of superintense radiation's interaction with matter, the production of X-ray radiation, and particle acceleration in the presence of a superstrong electromagnetic field. The Nike excimer laser system, based in the United States, is known for its exceptional power. It produces light pulses with an energy capacity of up to 5 kJ, lasting for 240 ns at full width half maximum (FWHM), and operates at a wavelength of 248 nm [1]. The output amplifier of this system has an aperture measuring 60 × 60 cm. The system was developed and used to address the issue of laser thermonuclear fusion. It is utilized in research involving the production of high-power nanosecond pulses and their interaction with a target. Japan developed the second largest excimer laser system, known as Super-Ashura [2]. The output amplifier has an aperture size of 61 cm. This device produces radiation pulses with an energy of up to 3.7 kilojoules (specifically, using KrF molecules) with a pulse duration of around 240 nanoseconds. The Super-Ashura system is utilized for conducting studies involving the production of high-energy nanosecond and picosecond pulses, as well as studying the effects of these pulses on matter. The Titania laser system, developed in England, utilizes KrF molecules to produce pulses with an energy of approximately

1 kilojoule and a pulse duration of around 150 nanoseconds [3]. The output amplifier has an aperture size of 42 cm. This apparatus is utilized for conducting research on the production of high-power picosecond and femtosecond pulses and their interaction with matter.

Russian institutes are involved in the development and production of high-powered excimer lasers and laser systems [4, 5, 6, 7, 8, 9, 10, 11, 12]. Specifically, the KrF laser system with a Garpun output unit, developed by the Institute of Physics, Russian Academy of Sciences, has an output aperture measuring 16 × 18 cm. This system produces radiation pulses with an energy of 80 J and a pulse duration of 100 ns [12].

The excimer lasers produced at the High-Current Electronics Institute (HCEI), Siberian Division, Russian Academy of Sciences, emit pulses with an energy exceeding 100 J [4, 5, 6, 8, 9, 10, 11]. The two most potent XeCl lasers, operating at a wavelength of 308 nm, have apertures measuring 40 cm and 60 cm. These lasers produce pulses with energy of 660 J and 1.9 kJ, with pulse lengths of around 350 ns and 250 ns, respectively.

The hybrid technique to femtosecond pulse amplification was developed in the late nineteenth century. The rare-gas-halide excimer media were employed to enhance femtosecond pulses, by excitation by either a high-voltage discharge or an electron beam. The laser in question utilizes the transitions between the excited B-state and the ground weakly bonded or weakly repulsive X-state of ArF, XeCl, XeF, or KrF excimer molecules [13]. However, B-X transitions have rather narrow gain bandwidths (broadest bandwidth of $\Delta\lambda \approx 2$ nm) and rather small saturation fluence ($\varepsilon_{sat} \approx 1$ mJ/cm^2). To realize high-output peak powers, (more than 1 TW) large apertures are required. Furthermore, a high gain of the amplifiers limits the temporal contrast of output radiation at the level of 10^2–10^3 [14]. The highest peak power of output radiation reached in traditional rare-gas-halide excimer amplifiers does not exceed ~ 4 TW [15].

A photodissociative-driven XeF(C-A) medium has a wide amplification band (~ 60 nm) in 475 nm range and a high saturation fluence of ~ 0.05 J/cm^2 unlike the traditional excimer molecules on the B-X transition [16]. At present, the development of ultra-high-power laser systems with a pulse duration of 10–100 fs is based mainly on near-infrared solid-state Ti:sapphire or parametric amplifiers. In these systems, positively chirped pulses, i.e., stretched in time (up to 0.5–1 ns) by linear frequency modulation [17], are amplified, and following its temporal recompression, the initial duration is realized. A pulse stretching allows avoiding nonlinear effects related to self-focusing. The pulses are normally stretched ~10^4 times, and for pulse recompression, a vacuum compressor based on diffraction gold-coated gratings is used.

An alternative approach to the design of multiterawatt and petawatt femtosecond laser systems has been developed at the Lebedev Physical Institute (Moscow, Russia) [18, 19], LP3 Laboratory of the Marseille University (Marseille, France) [20, 21, 22], and Institute of High Current Electronics (Tomsk, Russia) [23, 24, 25]. This approach is based on a solid-state femtosecond front-end and a photochemical XeF(C-A) boosting amplifier with a gaseous active medium [26, 27]. The hybrid (solid/gas) architecture offers an advantage in that the optical nonlinearity of gas is significantly lower than that of solids. As a result, the allowable stretching factor for femtosecond pulses is three orders of magnitude smaller compared to solid-state systems. This enables the amplification of picosecond chirped pulses and their subsequent compression using simpler and more efficient techniques. Negatively chirped pulses can be compressed in bulk glass by exploiting positive group velocity dispersion. Furthermore, the hybrid laser system functions inside the visible spectral range, which can be beneficial in some applications involving intense laser-matter interaction [28]. The active medium's gaseous nature facilitates the straightforward scaling of hybrid laser systems. The Institute of High Current Electronics (HCEI) SB RAS has created the THL-100, a multiterawatt hybrid laser system. This system is based on a photochemical-driven XeF(C-A) boosting amplifier with a 24 cm aperture.

This study presents three high-power excimer laser systems that were built at HCEI. These systems are based on wide-aperture lasers with different output laser beam sizes: 25 × 25 cm [6], 40 cm in diameter [8], and 25 cm [29, 30], respectively. The results acquired from the experiments conducted on these systems are reported.

2. MELS-4 K LASER SYSTEM

2.1. Experimental method and equipment

The MELS-4 k laser system is comprised of three main components: the master complex, preamplifier, and UFL-100 M amplifier [7]. The master complex comprises a pair of XeCl lasers that utilize electric-discharge for operation. One of them functions as the primary oscillator. The utilization of both lasers enables the manipulation of the characteristics of the emitted radiation. In particular, for single-pass lasing, the radiation energy is E = 15 mJ, the spectral line half width is $\Delta\nu$ = 0.01 cm^{-1}, and the pulse FWHM is 50 ns. In the case of injection locking, the parameters of the output beam are E = 100 mJ, $\Delta\nu$ = 0.01 cm^{-1}, and τ = 100 ns, while more than 50% of the radiation energy is concentrated inside the diffraction angle [31]. In the case of double-pass lasing with the phase conjugation, the parameters are E = 50 mJ, $\Delta\nu$ = 0.01–0.4 cm^{-1}, and τ = 30 ns, and the divergence is close to the diffraction-limited Q_d [32]. When the pulse is compressed to a pulse duration of 1–2 ns upon stimulated Brillouin scattering, the parameters are E = 10 mJ, $\Delta\nu$ = 0.01 cm^{-1}, and Q_d [33].

The preamplifier represents an electric-discharge lase with an active volume of 6x11x80 cm^3 (Figure 1). This laser consists of a metal housing that contains the dielectric laser chamber, capacitors with a total capacitance of 368 nF that are directly connected to the electrodes, and the X-ray source. A discharge gap and a storage capacitor (0.4 μF) are placed outside. The laser mixture Ne/Xe/HCl = 1000/10/1 is photo-ionized at a pressure of 2–4 atm. The storage capacitor is linked to the discharge gap, and 300 ns before the voltage across the electrodes reaches its highest point, the X-ray source is activated. The emission of radiation from this source triggers the discharge. The X-ray radiation is passed through a stainless steel grid that has a geometric transparency of 50%. The radiation dose within the laser chamber is approximately 25 milliroentgens (mR). The laser emits energy ranging from 6 to 10 joules with a pulse duration of 80 to 160 nanoseconds.

Figure 1. Photograph of the electric-discharge preamplifier.

The active volume of the primary amplifier is 62,500 cm^3. The gas is stimulated by two electron beams [6]. The electron accelerators are positioned vertically at the upper and lower ends of the laser chamber, which has a total internal volume of 360 liters. Within each accelerator, the vacuum diode and the high-voltage generator are contained within a unified metal enclosure. The cathode of the vacuum diode is securely attached to the final stage of the high-voltage generator. The laser's maximum energy in a plane-parallel cavity is 210 J, and the pulse duration is 250 ns. The windows of the laser chamber are inclined at an angle of 100 degrees with respect to the optical axis in the amplification mode. In order to initiate the master complex and secure its connection to the amplifiers, we utilize a high-voltage pulsed oscillator. The activation timing of the master oscillator and amplifier is regulated by manipulating the cable and employing artificial delay lines. When the radiation is increased in strength in the laser system, it meets the following requirement: the highest level of intensity in the initial pulse aligns with the highest level of power from the energy source in each of the amplifiers.

In order to ascertain the wave-front distortions of the amplified beam on the optical elements, we compute the optical path by considering the placements of the elements and the surface polish.

The wave-front distortion of the laser beam is limited to λ/4 for optical components with a diameter of up to 100 mm. When the diameter D exceeds 100 mm, the distortion does not exceed the wavelength λ. In order to determine the energy characteristics of the radiation, we employ numerical simulations to simulate the amplification modes. To achieve this objective, we utilize the one-dimensional model created at HCEI, which is founded on a set of nonstationary equations that describe the concentrations of excimer molecules and photon fluxes [7].

2.2. Calculated parameters of the amplified radiation

The short spontaneous lifetime (10–14 ns) of the upper laser level of excimer molecules and a high small signal gain g_0 (0.05–0.1 cm^{-1}) lead to a relatively high level of amplified spontaneous emission (ASE). The purpose of the calculations is to predict the output parameters of the laser system with allowance for the ASE in the optical schemes.

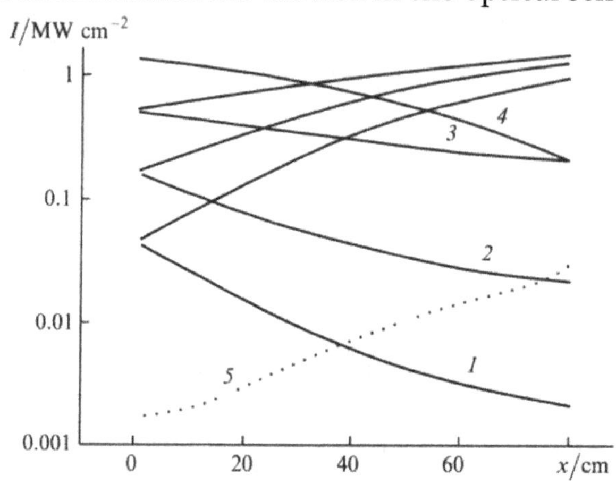

Figure 2. Intensity distributions of (1)–(4) the amplified signals and (5) the total ASE for (1)–(3) and (5) the double-pass and (4) single-pass amplification in the active medium of the preamplifier. The signal traveling from the right-hand side to the left-hand side is reflected by the mirror on the left-hand side. Curve 5 corresponding to the ASE is presented for the conditions of curve 1.

The output characteristics of the radiation are calculated for the preamplifier and the main amplifier where the ASE effect is maximal. All of the intensity distributions are presented for a time-constant pump power. Figure 2 demonstrates the radiation intensity distribution with respect to the length of the preamplifier active medium for various intensities of the input signal. The calculations are performed for the following initial parameters of the active medium: $g_0 = 0.08\ cm^{-1}$, $\alpha = 0.015\ cm^{-1}$, $\tau_{ef} = 2.5$ ns, and $\sigma = 4 \times 10^{-16}\ cm^2$, where α is absorption coefficient, τ_{ef} is the efficient life time of XeCl* molecule, and σ is the stimulated emission cross section the spontaneous lifetime of the XeCl* molecules is $\tau_s = 14$ ns. Based on the results of the calculations and the input pulse duration, we may conclude that an input energy of about 10 mJ is sufficient for the saturation of the amplifier at two passes.

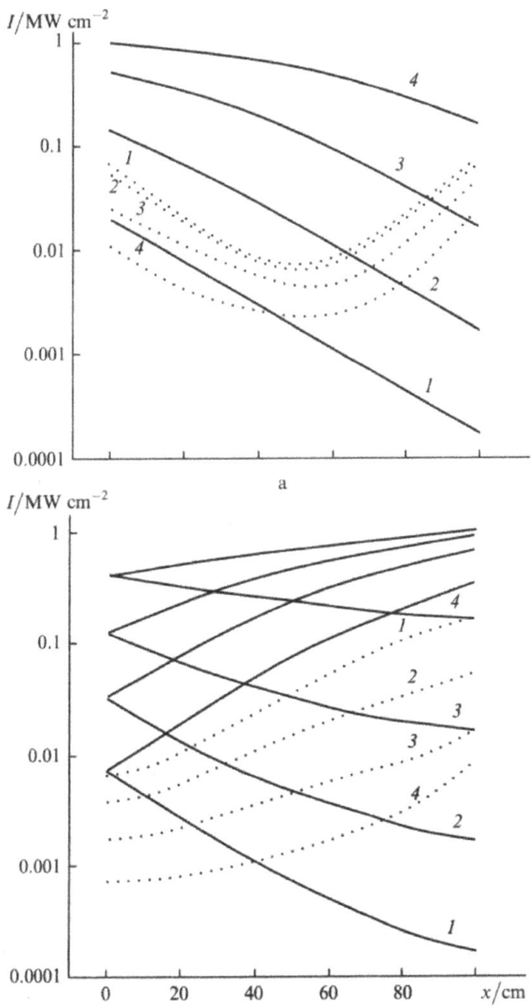

Figure 3. Intensity distributions of (solid lines 1–4) the amplified signals and (dashed lines 1–4) the total ASE for (a) one and (b) two passes in the active medium of the main amplifier.

In this case, the intensity of the amplified signal as about $2I_s$ (I_s is the saturation intensity). For input signals with an intensity of no less than 100 kW/cm^2, it is expedient to employ single-pass amplification. In general, the ASE effect is weak provided that the input intensity is more than 2 kW/cm^2 at the given parameters and sizes of the active medium.

For the main amplifier with a lower pump power, the parameters of the active medium are as follows: g_0 = 0.065 cm^{-1}, α = 0.0145 cm^{-1}, τ_{ef} = 3 ns, $\sigma = 4 \times 10^{-16}$ cm^2, and τ_s = 14 ns [7]. Figure 3 demonstrates the intensity distributions of the amplified radiation and ASE for the single-pass (a) and double-pass (b) configurations. It is seen that the ASE effect is significantly stronger than the effect in the preamplifier in spite of the comparable values of the product gL, where L is active medium length and g is the gain coefficient. For example, at a kilowatt level of the input signal, the ASE intensity at the output of the main preamplifier is close to the intensity of the amplified radiation, so that the latter cannot effectively make use of the population inversion. Note that the one-dimensional model slightly overestimates the signal-to-noise ratio (especially in the case of double-pass amplification). Therefore, the real ASE effect should be even stronger. Thus, the minimum intensity of the input signal of the main amplifier should be about 10 kW/cm^2. At a pulse duration of 80 ns, the corresponding energy density is about 1 mJ/cm^2. In this case, preference should be given to the double-pass configuration with regard to the energy of the amplified radiation.

For the computations, we utilize the nonstationary model and, as a result, we examine both the form of the intended signal and the ASE pulse shape [7]. The pulses at the input and output of the active medium may exhibit significant disparities. The discrepancy is contingent upon the disparity in the forms of the radiation and pump pulses, as well as the sharpness of the pulse edges and the duration of the active medium.

2.3. Experimental results

The experiments on the amplification of radiation in the laser systems are performed at a pulse duration of $\tau_{0.5}$ = 80 ns. The purpose of the experiments is to measure the energy characteristics of the amplifiers and to determine the wave-front distortions in the optical path. Figure 4 shows the experimental optical scheme. In the case under consideration, the amplified beam diverges owing to the presence of the negative lens 5. This makes it possible, on the one hand, to increase the output energy of the first stage and, on the other hand, to relatively easily match the apertures of the electric-discharge amplifiers. To match the beam size with the aperture of the output amplifier, we additionally employ telescope 6, which also provides for the spatial filtering of the radiation.

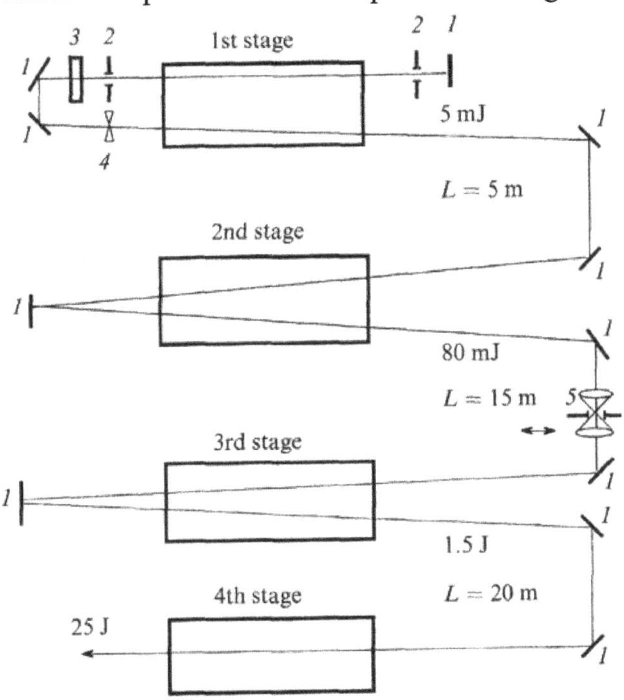

Figure 4. Optical scheme of the experiment: 1 and 4 totally reflecting mirrors, 2 pinhole with a diameter of 1.5 mm, 3 semitransparent mirror, 5 negative lens with F = - 80 cm, 6 telescope consisting of lenses with focal lengths of 50 and 100 cm (L is the optical path length in between the stages).

Table 1 demonstrates the measured output radiation energies and divergences for each of the amplification stages. Here, *A* is the beam size at the amplifier output; E_{in} and E_{out} are the input and output energies, respectively; and E_d is the energy inside the angle Q_d. The larger beam size in the preamplifier and amplifier corresponds to the presence of the telescope. The difference between the output and input energies of the stages is related to the loss at the turning mirrors and the loss resulting from the aperture matching.

Table 1. Parameters of the radiation amplified with an MELS-4 k laser system.

Parameters	First stage	Second stage	Third stage	Fourth stage
A, cm	Ø = 0.6	1.2 × 1.8	5 × 6	21 × 25
E_{in}, J	–	0.003	0.03	1.2
E_{out}, J	0.005	0.08	1.5	25
E_d/E_{out}	0.82	0.77	0.5	–

It is seen from Table 1 that the radiation divergence at the output of the first stage is close to the diffraction-limited divergence. Nevertheless, a worsening of the discharge homogeneity or an increase in the diameter of the beam amplified in it leads to a decrease in the energy concentrated inside the angle Q_d to a level of no greater than 50%. Using single-pass amplification in the passive part of the active medium of the first stage, we additionally increase the output energy of the master oscillator by a factor of about 20.

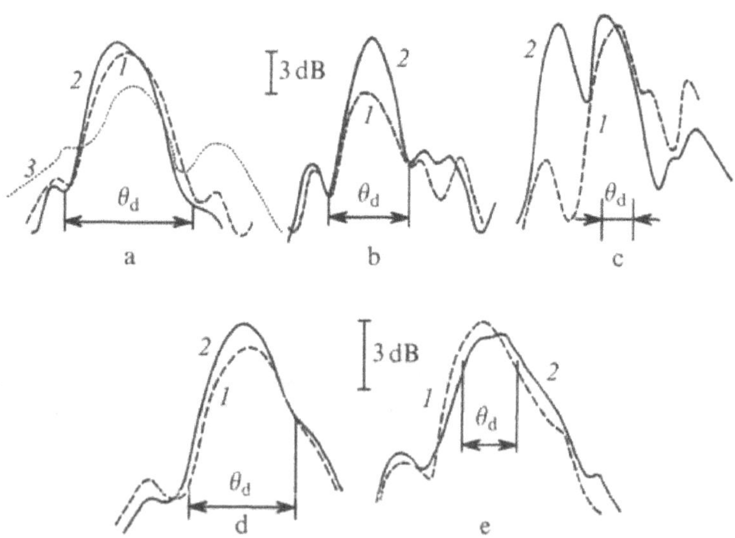

Figure 5. Densitograms of the focal spots for beam diameters of (a) 35, (b) and (c) 75, and (d) and (e) 150 mm: 1 original radiation, 2 amplified radiation, and 3 original radiation having passed through the amplifier a few seconds after its work.

During the second stage, an input energy of 3 millijoules is enough to fully saturate the active medium of the amplifier after two passes. The presence of the ASE is responsible for a slight rise in divergence after the second amplification step, compared to the divergence after the first stage. The active medium of the preamplifier does not contribute to the observed rise in divergence caused by distortions in the air and optical elements of the optical path.

The turbulent fluxes in the atmosphere hinder the measuring of the divergence at the output of the primary amplifier. Specifically, the location and arrangement of the focus point become unpredictable when the radiation from the master oscillator travels through the optical system without any amplification. The instability is highly influenced by the proximity of heat sources to the optical channel and the time elapsed after the amplifiers have been operated. This is the reason for the approximate value of the ratio E_d/E_{out} for a beam size of 10 × 12 cm. We may state that, in this case, nearly 50% of the energy is concentrated in an angle of 5×10^{-5} rad. This result is in agreement with the results of the alternative measurements in which the output radiation is focused by a lens with $F = 1.5$ m on the titanium foil with a thickness of 50 μm: a hole diameter of 100 μm corresponds to a divergence of about 6.5×10^{-5} rad.

To more thoroughly study the effect of the heterogeneities in the active medium of the main amplifier and the optical path, we amplify the radiation of the master oscillator at three beam diameters: 35, 75, and 150 mm, respectively. Figure 5 demonstrates the intensity distributions of the original radiation (the amplifier is switched off) and the amplified radiation measured in the far-field region. The first three panels correspond to the single-pass amplification in the active medium. It is seen that, for beam diameters of 35 and 75 mm, the divergence of the original and amplified radiation is close to the diffraction-limited divergence Q_d. At a beam diameter of 150 mm, the focal spot is broken into a few spots, so that the divergence is significantly higher than Q_d.

The likely cause of this phenomenon can be attributed to variations in air density along the optical path. In the specific scenario being discussed, the distance between lens 5 and the focus spot is around 25 meters. We modified the optical technique in order to minimize the potential impact of air. The novel approach involves use a telescope to enlarge the amplified radiation before it reaches the amplifier. The radiation is then amplified, reflected by a mirror, amplified again on its return, compressed using the same telescope, and finally detected. In terms of detection, the reflection mirror was somewhat out of alignment with respect to the optical axis of the system. Figure 5d and e display the densitograms of the focal locations for the new scheme. Observations indicate that the focal spot remains intact even when the beam diameter reaches 150 mm. The calculations demonstrate that the presence of spherical aberration in the telescope is responsible for a specific increase in size compared to the diffraction limit.

The time interval between the activations of the amplifier in the densitogram measurements is a minimum of 30 minutes. For smaller intervals of time between pauses, the spread of the initial radiation becomes greater, even when the beam has a diameter of 35 mm (Figure 5a). However, the distortions are not present when the light from the master laser enters the working medium of the amplifier immediately after the pump pulse ends.

3. LASER SYSTEM WITH AN OUTPUT APERTURE OF 40 CM

The laser system comprises five excimer lasers (Photon-1–Photon-5), along with the synchronization and ignition system, and the corresponding optical components. Three lasers utilize an electric discharge to stimulate the working combination. Electron-beam excitation is employed in two lasers. The laser radiation parameters are monitored in the experiments using conventional procedures and devices.

To measure the time and energy characteristics of the laser pulses, we employ an FEK-22 vacuum photodiode, TPI and IKT-2 N calorimeters, and an OPHIR calorimeter with an L30A-EX head. The signals are detected with Tektronix oscilloscopes.

3.1. Discharge lasers

Figure 6 displays the outer appearance of the initial laser device, known as Photon-1, as referenced in [34]. The electrodes of the discharge gap are positioned within the steel discharge chamber of the laser, which has a diameter of 35 cm. The electrode has a total length of 107 cm, with an active length of 102 cm. The interelectrode gap has a width of 4 cm. An insulator is used to cover a rectangular aperture on one side of the laser chamber. The components of the laser excitation system are positioned on the insulator surface and enclosed within a metal enclosure.

Figure 6. Photograph of the Photon-1 laser setup.

For the excitation, we employ an electric circuit with inductive energy storage and a semiconductor current breaker [35]. In this circuit, the high-voltage pulse with a short leading edge that is needed for discharge ignition is formed with a peaking capacitor $C = 3.2$ nF that is charged by the inductive energy storage with 12 SOS diodes [36]. The discharge preionization is realized upon the charging of the peaking capacitors owing to the radiation of 90 spark gaps that are uniformly distributed on the two sides of the anode. The main energy is deposited to the active medium at a low optimal voltage supplied by the storage capacitor (550 nF) that consists of K15–10 ceramic capacitors. The laser cavity consists of dielectric mirrors with reflectances of 98% and 7%. For the mixture Ne:Xe:HCl = 1520:10:1 with a pressure of 2 atm, the maximum laser energy is higher than 1.5 J at a pulse duration of 300 ns (FWHM) and a total efficiency of 1.35%. The low efficiency was due to the low content of HCl in the mixture, which was necessary to ignite the bulk discharge with large duration (~ 500 ns).

The two remaining electric-discharge lasers (Photon-2 and Photon-3) [37, 38] are virtually identical with respect to the design and the pumping scheme. Figure 7 demonstrates the exterior view of Photon-3. Each of these lasers consists of three units that contain the gas-discharge chamber with the pump generator and electric and pneumatic control panels. The chamber contains the vacuum diode of the soft X-ray source, the electrodes of the laser discharge gap, and the insulator of the high-voltage input. The electrodes are made of stainless steel. For the Photon-2 (Photon-3) laser, the interelectrode distance is 5.4 (9) cm and the electrode length is 80 (100) cm. The X-ray radiation is conveyed to the discharge gap via the electrode window, which is shielded by a titanium foil that is 80 micrometers thick. The anode of the discharge gap is linked to the pump generator, which is positioned outside the chamber, through an insulator equipped with metal studs. The design of the components that link the generator to the anode and the rearward wire-ways is optimized to minimize the inductance of the discharge circuit.

Figure 7. Photograph of the Photon-3 laser setup.

The vacuum diode of the X-ray source has a cylindrical housing that accommodates the anode and the cold cathode working upon explosive electron emission. The emission surface of the cathode consists of strips of foil-clad fiberglass plastic. The strips are attached to a separating grid that is covered with a 40-μm-thick titanium foil for the sealing of the vacuum diode and is used for the out coupling of the X-ray radiation. A tantalum foil serves as the anode of the vacuum diode. The vacuum diode is evacuated with an oil-diffusion pump to a residual pressure of about 10^{-4} torr.

A three-stage Arkad'ev-Marx generator with a fast capacitance of 15 nF serves as the power supply for the vacuum diode. The generator is connected to the vacuum diode with a KVI-120 high-voltage cable. A positive voltage pulse with an amplitude of 50–55 kV and a duration of 700 ns is fed to the anode. The X-ray doze in the cathode region of the discharge gap is 20–30 mR.

The storage capacitor, the switching unit, and the peaking capacitor are the main elements of the laser pump generator. The storage capacitor C_L consists of two (Photon-2) or three (Photon-3) FL-300 lines connected in parallel. The electric length of the line is 300 ns, its capacitance is 150 nF, and the impedance is 1 Ω. The pulsed charging of the line employs an IK-100 capacitor that is connected to the line with a KVI-120 cable. This capacitor can be charged to a voltage of 40–65 kV. The peaking capacitors C_2 = 4.9 (Photon-2) and 6.9 nF (Photon-3) that form the space discharge in the gap represent the batteries of KVI-3 ceramic capacitors (20 kV and 680 pF).

A rail-gap switch exhibits a low inductance. Its electrodes are made of stainless steel. For the Photon-2 (Photon-3) laser, the electrode length is 80 (100) cm and the interelectrode distance is 4 (6) mm. A starting electrode made up of foil leaves is located in the vicinity of the cathode along its entire length. The gaps are activated when a high-voltage pulse is fed to the starting electrode. The gap case represents a dielectric tube with an outer diameter of 65 mm. In the working mode, the gaps are filled with dry air at a pressure of 4–6.6 atm.

The high-voltage pulse needed for the activation of the spark gaps is produced by a high-voltage impulse generator based on a TGI-1-1000/25 thyratron. The artificial delay lines of the synchronization system provide for a sequential switching on of the pump generator and the X-ray source.

Plane-parallel plates made of fused quartz serve as the windows of the laser chamber. In the lasing mode, the laser cavity is formed by the external dielectric mirror with a reflectance of 98% at a wavelength of 308 nm and the window of the laser chamber. The laser mixture has a pressure of 3.5–4 atm and consists of the gases Ne, Xe, and HCl.

The free-running Photon-2 and Photon-3 lasers generate pulses with a duration of 250–300 ns (FWHM) and an energy of 3.5 and 10 J, respectively.

3.2. Electron beam–pumped lasers

Figure 8 displays an image of the fourth laser, known as Photon-4. The laser comprises the Arkad'ev-Marx generator (a high-voltage generator), a vacuum diode, a laser chamber, a gas-filling and evacuation system, and an electric control panel. The vacuum diode and associated power supply generator are housed together in a single enclosure. Therefore, the high-voltage generator achieves vacuum insulation for its high-voltage components. This design enables the reduction of inductance in the power-supply circuit of the vacuum diode, as well as the size and weight of the accelerator. The high-voltage generator consists of three parallel branches, which allow for the reduction of inductance and degradation of the gap electrodes. The spark-gap space in each branch is filled with a mixture of dry air and SF6. The capacitance of each stage within a single branch is 0.18 nanofarads. The accelerator is initiated by establishing a regulated distance and connecting it to a high-voltage cable that is linked to the high-voltage generator.

The gap is triggered by a high-voltage pulse from the thyratron generator. The vacuum-diode cathodes, measuring 110 cm in total length, are affixed to a holder located on the upper stage of the high-voltage generator. The laser chamber, functioning as the anode, is positioned within the vacuum diode, between the cathodes at the center. The laser chamber has an interior diameter of 25 cm. The chamber is connected to the diode housing across its entire length using a metal plate. The plate enhances the current flow and reduces the loss of the electron beam due to the influence of the self-magnetic field. Without the presence of the plate, if the current is just connected at the ends of the gas chamber, the loss can reach up to 50%. Velvet acts as the electron emitter on the cathodes of the vacuum diode. The distance between the anode and cathode is 7 cm. Therefore, four beams that converge towards the center are emitted from the diode and introduced into the laser chamber through a total of eight windows (two windows per row, with a combined length of 120 cm). Every airtight window is sealed shut using a 40-micrometer-thick titanium foil that is connected to a metal grid. The residual gas pressure in the enclosure housing the high-voltage generator and vacuum diode is 5×10^{-5} torr.

At a charging voltage of 85 kV, the voltage pulse formed by the generator across the diode has a duration of 1000 ns, an amplitude of 480 kV, and a total current of 74 kA. The electron beam generated in the diode provides for a relatively homogeneous excitation of the laser mixture.

The plane-parallel plates, which have a diameter of 300 mm and are made of fused quartz, serve as the windows of the laser chamber. In the lasing mode, the laser cavity is formed by the Al-coated external mirror and the window of the laser chamber.

Figure 8. Photograph of the Photon-4 laser setup.

The laser mixture consists of argon, xenon, and HCl. For the mixture Ar:Xe:HCl = 760:20:1 with a pressure of 2 atm, the radiation pulse energy amounts to 120 J at a charging voltage of 85 kV. The FWHM of the laser pulse is about 250 ns. Figure 9 shows the exterior view of the Photon-5 laser. The laser gas mixture is excited by a radially converging electron beam from six directions [8]. The beam is generated in a vacuum diode that contains 18 cathodes. The cathode profile is chosen in accordance with the results of the numerical calculations of the beam parameters using the original 2D code. The emitting cathode surface is made of carbotextim (graphite-fiber material with a resistivity of about $[5–50] \times 10^{-2}$ Ωm), and it is covered with velvet. The width of the emitting surface is 120 mm and the total area of the cathode-emitting surface is 0.95 m^2. The interelectrode gap between the emitting surface and the supporting structure of the outcoupling window is 6 cm. The supporting structure contains 18 windows (three windows in a row with a total length of 150 cm). The geometrical transparency of the beam outcoupling system is about 75%. The beam is outcoupled to the laser chamber through the Ti foil with a thickness of 40 μm. The laser cell diameter is 41 cm and the cell volume is 200 l.

The voltage pulse across the diode is generated using two paralleled linear transformers with the vacuum insulation of the secondary coil. Each of the transformers consists of ten stages, and each of the stages consists of eight IK-100-0.17 capacitors (100 kV, 17 μF, and 50 nH). The output power of the stage is about 12 GW. The voltage is fed to the diode (the cathode collector) with vacuum lines that serve as the secondary coils of the transformer. The diameter of the diode vacuum chamber is 131 cm and its length is 210 cm. The collector is suspended coaxially to the vacuum chamber on two springs in the upper part of the chamber. The vacuum chamber is evacuated using two AVDM-250 vacuum devices with liquid-nitrogen traps to a residual pressure in the diode of $(3–4) \times 10^{-5}$ torr.

Figure 9. Photograph of the Photon-5 laser setup.

At a charging voltage of 85 kV, the amplitude of the voltage pulse across the vacuum diode amounts to 550 kV. In this case, the total current (the sum of two currents) is 320 kA, and the energy transferred from the transformer to the diode is 87 kJ. At a charging voltage of 80 kV, the voltage across the vacuum diode is 440 kV, the total current is 290 kA, and the energy transferred to the diode is 78 kJ. The energy input to the gas increases when the pressure increases to 2.5 atm. Then, the energy input remains virtually unchanged when the pressure increases to 3.5 atm. The maximum energy input to the gas from the electron beam is about 19 kJ. The efficiency of the energy transfer from the primary storage to the gas is about 19%. This result is close to the value realized in conventional accelerators with water lines [1, 2, 3].

Plane-parallel plates with a diameter of 400 mm made of fused quartz serve as the windows of the laser chamber. This laser is tested in lasing mode with the cavity formed by the Al-coated plane mirror and the laser chamber window. For the mixture Ar:Xe:HCl = 760:20:1 with a pressure of 2 atm, the radiation pulse energy amounts to 660 J at a charging voltage of 85 kV [8]. The radiation pulse FWHM is about 350 ns. The heterogeneity of the radiation energy density distribution over the cross section of the laser beam is less than 10%.

3.3. Laser system

Figure 10 depicts the block diagram of the synchronization and starting system utilized for synchronizing the setups. The laser system is operated using a personal computer (PC) that is connected to a synchronization pulse generator. This generator produces voltage pulses with an amplitude of 600 V and allows for the adjustment of the time delay between consecutive pulses. The pulses are employed to initiate four thyratron generators and the magnetic-bias generator of the Photon-5 laser. The final generator generates two bias pulses which are then supplied to the linear transformers in order to pre-magnetize the transformer cores in the correct orientation. The thyratron generators generate negative voltage pulses with an approximate amplitude of 20 kilovolts. The pulses are sent to the spark gaps of the Photon-1, Photon-2, and Photon-3 lasers, as well as to the inputs of the generators of the Photon-4 and Photon-5 lasers. Both generators generate negative pulses with an amplitude of around 85 kV. These pulses are utilized to activate the rail gaps of the Photon-2 and Photon-3 lasers, as well as initiate the Photon-4 laser. The pulses are also transmitted to the inputs of the trigger generators G-1 and G-2. The trigger generators generate 40 negative pulses with an amplitude of 85 kV, which then activate the gaps in the transformer stages of the Photon-5 laser. Furthermore, the PC governs the activation and deactivation of the capacitive storage charging in all lasers, taking into account the specific charging durations, to ensure that the storages are charged synchronously.

Thus, at the first stage of the laser system's operation, the storage capacitors of all of the lasers are charged. Then, the starting pulse is fed to the magnetic-bias generator of the Photon-5 laser and a certain sequence of the starting pulses is fed to the thyratron generators of the Photon-1, Photon-2, Photon-3, and Photon-4 lasers. Thus, we activate the system as a whole.

In the laser system, the Photon-2, Photon-3, Photon-4, and Photon-5 lasers start working as slave amplifiers. In this case, the windows of all of the laser chambers are tilted at angles providing for the absence of feedback. The Photon-1 laser serves as the master oscillator [34]. Its optical scheme makes it possible to generate a high-quality beam in a certain part of the active volume and to amplify this beam in the remaining part. To improve the spatial structure of the radiation, we employ two pinholes with a diameter of 2 mm. In this case, the Fresnel number is $N \sim 2$ for a cavity with a length of 1.5 m. The spectral selection is realized with an auto-collimation diffraction grating (1800 mm^{-1}). The feedback in the cavity is maintained via the first diffraction order. To decrease the contribution of the noise component to the output radiation, we outcouple the laser beam through a semitransparent mirror with a reflectance of R = 30%. Then, the low-power high-quality radiation of the master laser is additionally amplified at two passes in the same active medium, so that the output beam diameter increases to 7 mm. The output pulse of the Photon-1 laser has an energy of 50 mJ, a duration of 250 ns, and a spectral line width of 0.9 cm^{-1}. The divergence of the laser beam, which contains 80% of the energy (0.13 mrad), is greater than the diffraction-limited divergence by a factor of 1.2.

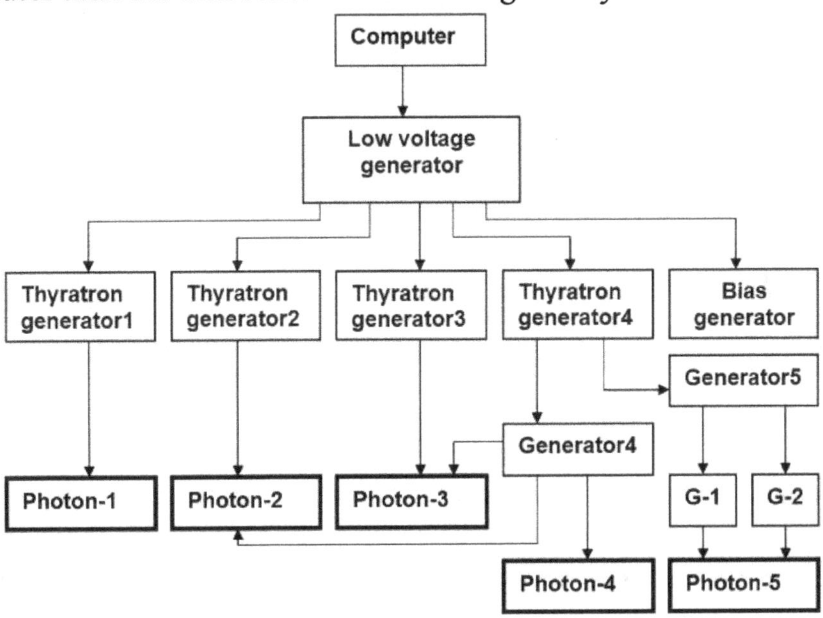

Figure 10. Block diagram of the starting and synchronization system.

This beam is expanded using a lens telescope with a magnification of M = 1.5 to match the beam diameter with the sizes of the active media of the Photon-2 and Photon-3 lasers. The beam is amplified at three passes in the Photon-2 laser and one pass in the Photon-3 laser. The output beam diameters of these lasers are 3 and 6 cm, respectively. For further matching of the beam diameter with the sizes of the active media of the

Photon-4 and Photon-5 lasers, we employ a lens telescope with a fivefold magnification. After the beam expansion, the radiation is amplified at one pass in the active medium of the Photon-4 laser and at one or two passes in the active medium of the Photon-5 laser.

Table 2 summarizes the experimental results for the laser system. The maximum energy (330 J) is obtained in the case of the single-pass amplification in the Photon-5 laser when the ASE flux and the absorption in the active medium are minimized. In the case of double pass amplification, both the absorption and the ASE contribution increase (the ASE intensity increases owing to the reflection from the rear mirror with $R = 99\%$). This leads to a decrease in the energy of the amplified radiation to a level of 250 J.

Table 2. Parameters of the radiation of the photon lasers.

Lasers	Operating mode	Energy, J	Pulse duration, ns	Line width, cm^{-1}	Divergence, μrad
Photon-1	Laser master oscillator	1.50.05	300 200–250	– 0.9	– 130
Photon-2	Laser amplifier	3.50.5	250200–250	– 0.9	– 60
Photon-3	Laser amplifier	105	300200–250	– 0.9	– 37
Photon-4	Laser amplifier	12040	250200–250	– –	– –
Photon-5	Laser amplifier	660250, 330	350200–250	– –	– –

The spectral and spatial parameters of the radiation are measured only for the first three setups (Photon-1, Photon-2, and Photon-3). The divergence is measured using the spot size in the focal plane of a lens with a focal length of $F = 13.5$ m. The line width is measured with the Fabry-Perot interferometer. In both cases, the radiation intensity distribution is detected by a CCD array. The spectral measurements show that the line width remains unchanged (0.9 cm^{-1}) after the amplification. In general, the divergence of the amplified radiation decreases with an increase in the beam diameter. However, the divergence is slightly greater than the diffraction-limited divergence due to the distortions in the active medium and the optical path.

Figure 11 shows the dependence of the output energy of the main amplifier on the input energy upon the double-pass amplification. The input energy is measured when one of the amplifiers (Photon-3 or Photon-4) is switched off. It is seen that the Photon-5 amplifier is saturated only when all of the amplifiers are switched on. In this case, the gain is about 10. The laser beam spot on photosensitive paper exhibits a relatively homogeneous distribution and a minor contribution of the diffraction rings related to the heterogeneities in the optical path. This indicates a high spatial coherence of the output radiation.

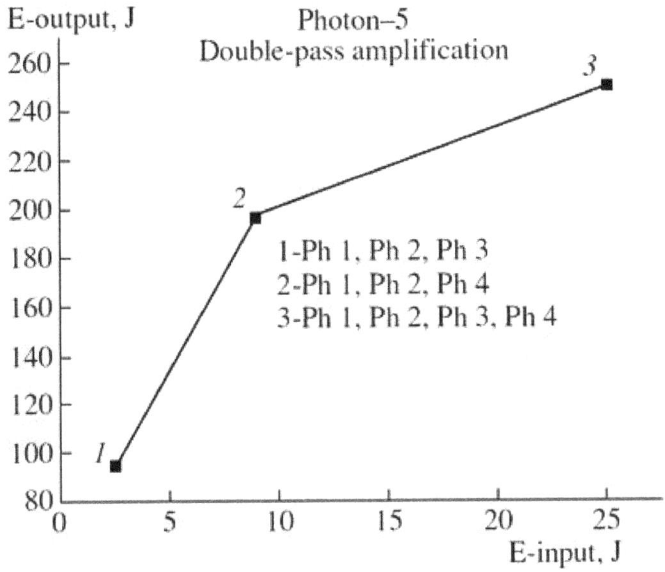

Figure 11. Plot of the output radiation energy vs. the input signal energy for the double-pass amplification in the active medium of the Photon-5 laser.

4. LASER SYSTEM THL-100

4.1. Femtosecond pulse generator

The front-end of the laser system comprises a femtosecond Ti:sapphire master oscillator, regenerative and multipass amplifiers, a pulse stretcher and compressor, and a second harmonic generator. It functions with a pulse repetition rate of 10 Hz and operates in a mode where just a single pulse is emitted at a time. The front-end's output energy can reach a maximum of 5 millijoules at a wavelength of 475 nanometers. The pulse duration is limited to 50 femtoseconds.

4.2. Photodissociation excimer XeF(C-A) amplifier

The Figures 12 and 13 display the XeF(C-A) amplifier and its schematic cross section, respectively. The setup comprises of two high-voltage pulse generators, a vacuum chamber containing six electron cathodes, a xenon gas chamber, and a laser cell. The laser cell used for multipass laser beam amplification contains a mirror unit consisting of 32 mirrors.

Two high-voltage pulse generators are line transformers, which have twelve transformer stages. Each stage has 320 nF capacitors and multigap spark switch. A secondary turn of the transformer is a vacuum coaxial line, which is connected with an e-beam diode. The diode has six explosive emission cathodes (flock-coated metal). The anode of vacuum diode is 45 cm-dia gas chamber closed by 40-μm-thick titanium foil (Figure 14). The vacuum diode forms six radially converging electron beams, each of 15 × 100 cm^2 area, that are injected into gas chamber filled by xenon at a 3 atm pressure. The capacitors of each stage can be charged up to 90–100 kV.

Figure 12. Photo of the XeF(C-A) amplifier.

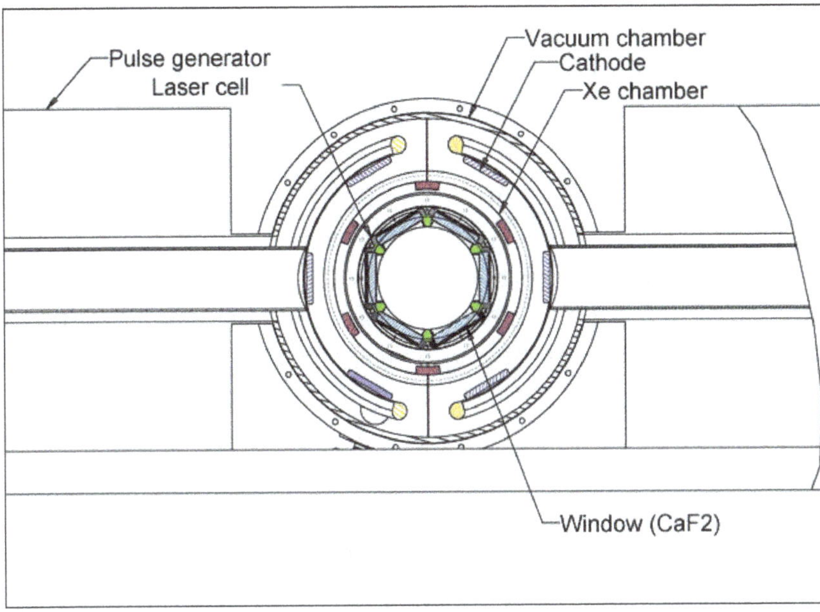

Figure 13. Schematic of the pump circuit of the XeF(C-A) amplifier active medium.

Figure 14. Gas chamber of the XeF(C-A) amplifier with windows for electron beam injection.

At charge voltage of 95 kV, the energy stored in the capacitor of a line transformer is 34.6 kJ, the energy of electron beam in the vacuum diode is 21 kJ, and the energy delivered to the gas chamber is 7 kJ. E-beam in the vacuum diode has following parameters: peak voltage of 550 kV, peak current of 200 kA, and current pulse duration of 150 ns (FWHM).

The electron beam is injected into the gas chamber and one almost completely is absorbed by xenon at its thickness of 8 cm. An efficiency of the electron beam energy converting to VUV radiation of Xe_2^* excimers is 30–40% [27].

The laser cell, depicted in Figure 15, is positioned within the gas chamber along its central axis. It is filled with a gas mixture consisting of XeF_2 and high-purity nitrogen. The XeF_2 gas is present at a partial pressure ranging from 0.2 to 0.4 torr, while the high-purity nitrogen is present at a partial pressure ranging from 190 to 380 torr. The amplifier on XeF(C-A) molecules is stimulated by VUV radiation (172 nm) by the photodissociation of XeF_2 molecules, resulting in the production of XeF(B) molecules. The XeF(C) state of the laser (C-A) transition is created as a result of the relaxation of XeF(B) molecules by collisions with a buffer gas composed of N_2. The 110 cm long active medium is circulated through arrays of 54 CaF2 windows, each measuring 12 × 12 cm^2, which are positioned on the side walls of the hexahedral laser cell. These windows are used to direct the electron beams into the gas chamber. This guarantees the optimal alignment of the pumping source with the active medium. The windows are hermetically sealed with Viton gaskets. The hexagon laser cell possesses a transparent opening with a diameter of 25 cm. The laser cells are hermetically sealed by the end flanges using fused silica windows with a diameter of 30 cm.

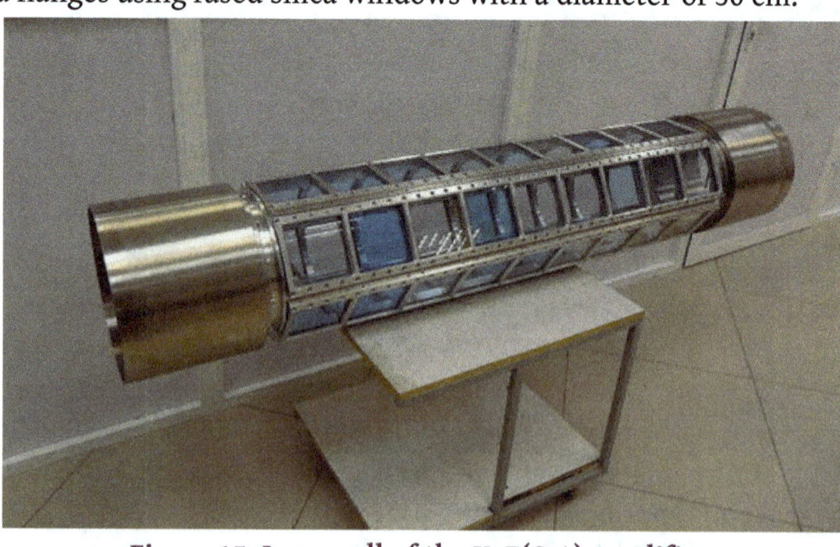

Figure 15. Laser cell of the XeF(C-A) amplifier.

A high-purity xenon (99.9997%) was supplied to the gas chamber to provide maximum efficiency of the e-beam to VUV radiation conversion. The chamber preliminary evacuated to a pressure of 10^{-4} torr. During operation of the amplifier, the intensity of xenon VUV radiation gradually decreased due to outgassing from the stainless steel walls of the gas chamber, exposed to the electron beam. Xenon was circulated through a Sircal MP-2000 purifier to support the gas purity recovery.

In our experiments, the mixture in the laser cell was replaced after each shot, because repeated pumping of the mixture decreased the output energy by 20–30%.

4.3. Measurement procedures

The front-end laser beam was amplified in the XeF(C-A) amplifier using a multipass optical system consisting of 33 passes. This scheme was created by using 32 spherical mirrors that increased in diameter. The mirrors, which had a reflectivity of 99.7%, were securely attached to the edges of the laser cell's intracell flanges. A laser beam of a divergent nature is introduced into the laser cell, causing it to increase in size from 2 cm at the entrance to 6 cm after 32 cycles. While traversing the mirrors, the beam completes two full revolutions within the laser cell. The second-to-last convex mirror redirects the beam towards a flat mirror with a diameter of 10 cm, which is positioned on the optical axis of the laser cell. A reflected beam travels down the optical axis and exits the laser cell with a diameter of 12 cm.

The laser pulse in the XeF(C-A) amplifier was initially elongated to 1 picosecond using a prism stretcher with negative group-velocity dispersion. The amplified laser pulse, which had a negative chirp, was enlarged to a length of 20 cm. It was then focused and compressed using three fused silica plates, each with a thickness of 4 cm, at a Brewster angle on two passes. The compressor experienced an energy loss of around 2%. To measure the duration of the amplified pulse, we attenuated the central section of the beam, which had a diameter of 90 mm, using two wedges. We then concentrated this attenuated beam using a spherical mirror with a focal length of 12 m. The focused beam was directed towards a 0.25 mm-diameter aperture, which was positioned in front of an ASF-20-480 single-shot autocorrelator. The sech2 function was included in the autocorrelator to accurately model the temporal intensity profile.

The TPI-2-7 calorimeter was used to measure the energy of the e-beams in the converter and VUV radiation in the laser chamber. The energy output of the XeF(C-A) amplifier was measured using an OPHIR energy meter. The output beam was attenuated using a fused silica wedge and focused to a spot size of 2.5 cm. The portion of the beam that traversed the wedge plate was utilized to capture a picture of the laser beam on a photographic paper. The tiny signal gain of the active medium was determined by measuring the output of a Sapphire-488 CW semiconductor laser after the probe beam passed through the active medium four times. This laser generates light with a wavelength of 488 nm, which aligns with the highest point of amplification in the XeF(C-A) transition.

4.4. Experimental results

First of all, the VUV radiation energy transmitted through the CaF_2 windows into the laser cell was measured. Its value was 240–260 J. In view of the quantum efficiency of laser transition and 100% quantum yield of XeF(C) state production, the integral value of the energy stored in the active medium is ~ 90 J. In actuality, the actual lifetime of the XeF(C) state, which is determined by radiative decay and quenching, is much shorter than the pump pulse width. This makes the maximum current value of the energy stored on the XeF(C-A) transition 10 times less than the integral value.

The time profile of the small-signal gain near the CaF_2 windows is shown in Figure 16. These results correlate well with those found in the experiments on femtosecond pulse amplification. As can be seen in the Figure 16, the maximum gain is 0.004 cm^{-1} and the FWHM of the amplified signal is ~ 200 ns. The amplification of picosecond pulses was performed within the time interval of 146 ns (33 passes) close to the gain profile maximum.

The small-signal gain distribution over the active medium cross section, measured at different XeF_2 vapor pressures, is shown in Figure 17. These measurements show that decreasing the vapor pressure enhances the gain distribution uniformity, but at the same time, it greatly decreases the gain near the pump windows. Final optimization of the gas mixture composition was made by achieving the maximum output energy of the amplified picosecond pulse in the real multipass amplification scheme. The maximum output energy was obtained for a XeF_2 vapor pressure of 0.2–0.25 torr.

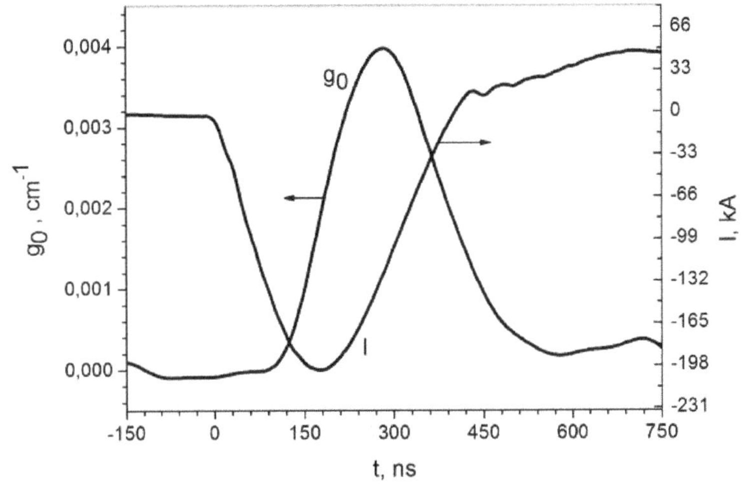

Figure 16. Time profile of the e-beam current in the diode and gain measured near the CaF_2 windows with the continuous laser at 488 nm for a XeF_2 vapor pressure of 0.25 torr.

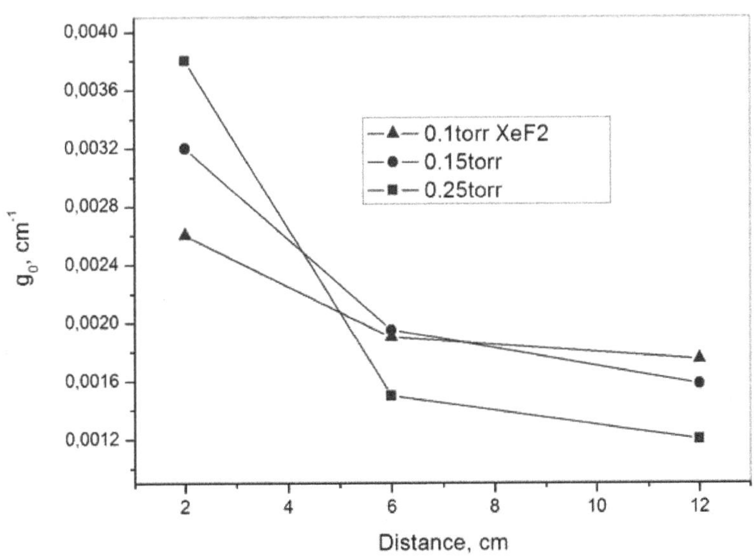

Figure 17. Small signal gain distribution from the window toward the center of the laser cell for different XeF_2 vapor pressures. The nitrogen pressure is 190 torr.

In our experiments, the seed pulse energy varied from 0.04 to 2 mJ. At an input energy of 0.04 mJ, the gain of active medium is far from saturation. The total gain factor was 2.5×10^3 by 100 mJ output energy. The total gain factor was reduced down to 5×10^2 under near-saturation conditions at a seed pulse energy of 2 mJ. In this case, the output energy was 1 J. The shot-to-shot fluctuations in the output pulse energy were within 10%. An imprint of the output laser beam on a photographic paper is shown in Figure 18.

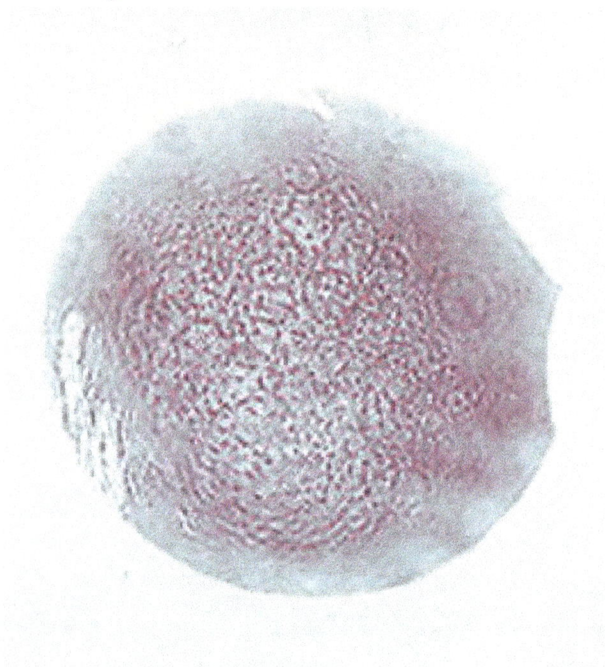

Figure 18. Autograph of the output laser beam.

The ASE power of the XeF(C-A) amplifier with seed pulse blocked was measured for an estimation of the contrast ratio. It was 1 W within an angle of 0.2 mrad, which is close to the output beam divergence angle. Thus, the temporal contrast ratio between the output laser peak and background ASE was ~ 10^{13}. It means that the real contrast of output beam is determined by the seed pulse temporal contrast, which is usually ~ 10^{10} at the second harmonic frequency [39].

After compression, the pulse duration was measured both with and without amplification in active medium. In both cases, the pulse duration was 50–60 fs inferred assuming $sech^2$ pulse-shape. The output beam energy was 0.5–0.7 J in these experiments. The autocorrelation function for the compressed pulse at 0.7 J output energy, which corresponds to 50 fs pulse duration, is shown in Figure 19. This result gives the peak power ~ 14 TW. According to literature data, the highest power of femtosecond pulses in the visible spectrum was attained earlier upon nonlinear conversion to the second harmonic in a Ti:sapphire laser [39] and upon direct amplification of 250-fs pulses in an electron beam-excited XeF(C-A) amplifier [40]. These values (4 TW and 1 TW, correspondingly) are far below the power attained in the present work.

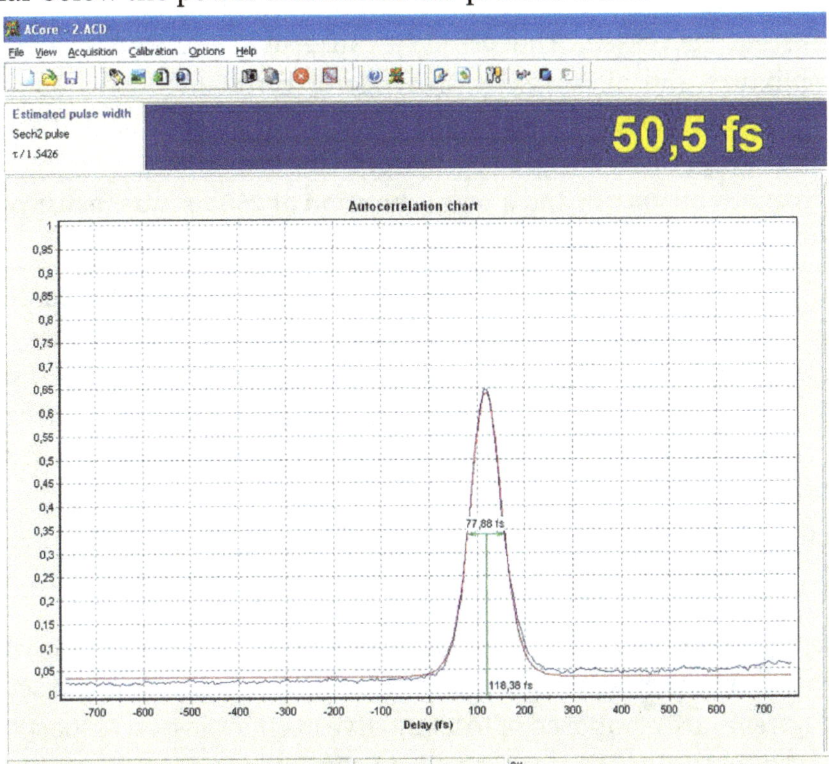

Figure 19. Pulse duration of the laser beam with energy of 0.7 J.

5. CONCISE SUMMARY

Three high-performance excimer laser systems developed at HCEI were showcased. An analysis was conducted on the designs of compact, high-powered lasers that are energized by electric discharge and electron beam. Research has demonstrated that electron beam–pumped lasers exhibit higher efficiency when used for lengthy laser pulse durations, specifically those lasting 300 ns or longer. A study presented the outcomes of producing a radiation with high quality and high strength in the ultraviolet (UV) and visible range.

Our experimental data confirms that it is possible to amplify diffraction-limited laser beams with a large cross section in a gas laser system. Experimental results demonstrated that the divergence of the intensified ultraviolet (UV) beam closely approximated that of a diffraction-limited beam, provided that its diameter did not exceed 5 cm. However, the divergence became even greater as the diameter was further increased. The primary cause of the distortions in the laser mixture and air was the presence of turbulent fluxes affecting the wave-front. A laser beam with a minimal divergence of 10 µrad was achieved. The pulse energy of the amplified ultraviolet (UV) light, with a wavelength of 308 nm and a pulse duration of 250 ns, was measured to be 330 J.

In the initial pilot studies of amplifying down-chirped 1 ps pulses in the THL-100 femtosecond hybrid laser system, an output energy of 1 J was achieved. The amplified pulses were recompressed in bulk fused silica to their initial duration of 50 fs, achieving an output energy of 0.7 J. This indicates that a remarkable level of power, approximately 14 terawatts, has been achieved in the visible spectrum, setting a new record.

AT A GLANCE

Three high-power excimer laser systems are described, each of which has an output laser beam aperture of 25 and 40 cm. The first and second laser systems are composed of four and five excimer lasers, respectively. The third system comprises a XeF(C-A) amplifier and a Ti:Sa front end. The experimental results of the generation of high-quality and high-power laser pulses are presented. Laser beams were produced with a maximal power of 14 TW (450 nm, 50 fs) and a pulse energy of up to 330 J (308 nm, 250 ns).

REFERENCES

1. Obenschain SP, Bodner SE, Colombant D, Gerber K, Lehmberg RH, McLean EA, Mostivych AN, Pronko MS, Pawley CJ, Schmitt

AJ, Sethian JD, Serlin V, Stamper JA, Sullivan CA. The Nike KrF laser facility: Performance and initial target experiments. Physics of Plasmas. 1996;3:2098-2107. Available from: DOI: https://doi.org/10.1063/1.871661

2. Owadano Y, Okudo I, Matsumoto Y, Tanimoto M, Tomie T, Koyama K, Yano M. Development of a high-power laser system, ASURA. Laser and Particle Beams. 1989;7:383-392. DOI: https://doi.org/10.1017/S0263034600007345

3. Dival EJ, Edwards CB, Hirst GJ, Hooker CJ, Kidd AK, Lister JMD, Mathumo R, Ross IN, Shaw MJ, Toner WT, Visser AP, Wyborn BE. Titania – A 10^{20} Wcm^{-2} ultraviolet laser. Journal Modern Optics. 1996;43:1025-1033

4. Bychkov YI, Ivanov NG, Losev VF, Mesyats GA. XeCl-laser with 150 J laser energy. Soviet Technical Physics Letters. 1988;14:253-256

5. Abdullin EN, Gorbachev SI, Efremov AM, Koval'chuk BM, Loginov SV, Skakun VS, Tarasenko VF, Tolkachev VS, Fedenev AV, Fomin EA, Shchanin PM. High-power compact laser with λ = 308 and 249 nm pumped by a radially converging electron beam. Quantum Electronics. 1993;23:564-567. Available from: https://doi.org/10.1070/QE1993v023n 07ABEH 003124

6. Ivanov NG, Losev VF, Naats EI, Ryzhov VV, Turchanovskii IY, Yastremskii AG. XeCl laser with an output energy 200 J. Quantum Electronics. 1997;27:670-672. Available from: https://doi.org/10.1070/QE1997v027n08ABE H001029

7. Ivanov NG, Losev VF, Panchenko YN, Yastremskii AG. XeCl laser system with a 25 cm × 25 cm output aperture. Quantum Electronics. 1999;29:852-856. DOI: https://doi.org/ 10.1070/QE1999v029n10ABEH0015 89

8. Abdullin EN, Grishin DM, Gubanov VP, Zorin VB, Kim AA, Koval'chuk BM, Kumpyak EV, Liu J, Morozov AV, Skakun VS, Stepchenko AS, Tarasenko VF, Tolkachev VS, Tsoi NV, Shchanin PM. A 650-J XeCl laser. Quantum Electronics. 2004;34:199-202. Available from: https://doi.org/10.1070/QE2004v034n03ABEH00 2610

9. Mesyats GA, Osipov VV, Tarasenko VF. Pulse Gas Lasers (Nauka, Moscow, 1991). Washington: Opt. Eng. Press; 1995. 272 p

10. Kovalchuk BM, Losev VF, Mesyats GA, Tarasenko VF. Electron-beam-pumped high-power wide-aperture exciplex lasers and laser systems. Russian Physics Journal. 2000;43:352-357. DOI: 10.1007/BF025 08516

11. Bugaev SP, Abdullin EN, Zorin VB, Koval'chuk BM, Loginov SV, Mesyats GA, Skakun VS,Tarasenko VF, Tolkachev VS, Shchanin PM. A 2-kJ wide-aperture XeCl laser. Quantum Electronics. 2004;34:801-804. DOI: https://doi.org/10.1070/QE2004v 034n09 ABEH002687

12. Zvorykin VD, Arlantsev SV, Bakaev VG, Rantsev OV, Sergeev PB, Sychugov GV, Tserkovnikov AY. Transport of electron beams and stability of optical windows in high-power e-beam-pumped krypton fluoride lasers. Laser Particle Beams. 2001;19:609-622. DOI: 10.1017/S0263034601194085

13. Rhodes CK. Excimer Lasers. Berlin-Heidelberg: Springer-Verlag; 1984. 245 p

14. McIntyre IA, Rhodes CK. High power ultrafast excimer lasers. Journal of Applied Physics. 1991;69:121-139. DOI: 10.1063/1.347665

15. Watanabe M, Endoh A, Sarukura N, Watanabe S. Subpicosecond UV pulse generation for a multiterawatt KrF laser. Applied Physics B. 1989;48:417-420. DOI: 10.1007/BF00694542

16. Bischel WK, Eckstrom DJ, Walker HC, Tilton RA. Photolytically pumped XeF(C-A) laser studies. Journal of Applied Physics. 1981;81:4429-4434. DOI: 10.1063/1.329357

17. Strickland D, Mourou GA. Compression of amplified chirped optical pulses. Optics Communications. 1985;56:219-221. DOI: 10.1016/0030-4018(85)90151-8

18. Mikheev LD. On the possibility of amplification of a femtosecond pulse up to the energy 1 kJ. Laser and Particle Beams. 1992;10:473-478. DOI: 10.1017/S02630346000067

19. Aristov AI, Grudtsin YV, Zubarev IG, Ivanov NG, Konyashenko AV, Krokhin ON, Losev VF, Mavritsky AO, Mamaev SB, Mesyats GA, Mikheev LD, Panchenko YN, Rastvortseva AA, Ratakhin NA, Sentis

M, Starodub AN, Tenyakov SY, Uteza O, Tcheremiskin VI, Jalovoi VI. Hybrid femtosecond laser system based on photochemical XeF(C-A) amplifier with an aperture of 12 cm. Atmospheric and Oceanic Optics. 2009;22:1029-1034

20. Malinovskii GY, Mamaev SB, Mikheev LD, Moskalev TJ, Sentis ML, Tcheremiskine VI, Yalovoi VI. Numerical simulation of the active medium and investigation of the pump source for the development of a photochemical XeF(C-A) amplifier of femtosecond optical pulses. Quantum Electronics. 2001;31:617-622. DOI: 10.1070/QE2001v031n07ABEH002014

21. Clady R, Coustillier G, Gastaud M, Sentis M, Spiga P, Tcheremiskine V, Uteza O, Mikheev LD,Mislavskii V, Chambaret JP, Chȇriaux G. Architecture of a blue high contrast multiterawatt ultrashort laser. Applied Physics B-Lasers and Optics. 2006;82:347-358. DOI: 10.1007/s00340-005-2081-1

22. Tcheremiskine V, Uteza O, Aristov A, Sentis M, Mikheev L. Optical sources based on a multichannel surface discharge and their application to pump photolytically driven femtosecond XeF(C-A) amplifier. Proceedings of SPIE. 2008;7005:70051K. DOI: 10.1117/12.782779

23. Alexeev SV, Ivanov NG, Kovalchuk BM, Losev VF, Mesyats GA, Mikheev LD, Panchenko YN, Ratakhin NA, Yastremsky AG. Hybrid femtosecond laser system THL-100 on the base of XeF(C-A) amplifier. Atmospheric and Oceanic Optics. 2012;25:221-225

24. Losev V, Alekseev S, Ivanov N, Kovalchuk B, Mikheev L, Mesyats G, Panchenko Y, Puchikin A, Ratakhin N, Yastremsky A. Development of a 100-terawatt hybrid femtosecond laser system. Proceedings of SPIE. 2011;7393:739317. DOI: 10.1117/12.879123

25. Losev V, Alekseev S, Ivanov N, Kovalchuk B, Mikheev L, Mesyats G, Panchenko Y, Ratakhin N, Yastremsky A. Development of a 100-terawatt hybrid femtosecond laser system on base of photochemically driven XeF(C-A) amplifier. Proceedigs of . SPIE. 2012;1461:116-119. DOI: 10.1063/1.4736773

26. Mikheev LD, Stavrovskii DB, Zuev VS. Photodissociation XeF laser operating in the visible and UV regions. Journal of Russian Laser Research. 1995;16:427-475. DOI: 10.1007/BF02581226

27. Eckstrom DJ, Walker HC. Multijoul performance of the photolytically pumped XeF(C-A) laser. IEEE Journal of Quantum Electronics. 1982;QE18:176-181. DOI: 10.1109/JQE.1982.1071517

28. Mikheev LD, Tcheremiskine VI, Uteza OP, Sentis ML. Photochemical gas lasers and hybrid (solid/gas) blue-green femtosecond systems. Progress in Quantum Electronics. 2012;36:98-142. DOI: 10.1016/j.pquantelec.2012.03.004

29. Alekseev SV, Aristov AI, Ivanov NG, Kovalchuk BM, Losev VF, Mesyats GA, Mikheev LD, Panchenko YN, Ratakhin NA. Multiterawatt femtosecond laser system in the visible with photochemically driven XeF(C-A) boosting amplifier. Laser and Particle Beams. 2013;31:17-21. DOI: 10.1017/S0263034612000 870

30. Alekseev SV, Aristov AI, Grudtsyn YV, Ivanov NG, Koval'chuk BM, Losev VF, Mamaev SB, Mesyats GA, Mikheev LD, Panchenko YN, Polivin AV, Stepanov SG, Ratakhin NA, Yalovoi VI, Yastremskii AG. Visible-range hybrid femtosecond systems based on a XeF(C-A) amplifier: State of the art and prospects. Quantum Electronics. 2013;43:190-200. DOI: 10.1070/QE2013v043n03ABEH015096

31. Ivanov NG, Kovalenko SE, Losev VF, Panchenko YN. The laser system on XeCl molecules with high spectral brightness. Atmospheric and Oceanic Optics. 1995;8:1590-1594

32. Byskov YI, Losev VF, Panchenko YN. Experimental investigation of the efficiency of phase conjugation of a XeCl laser-beam by stimulated brillouin-scattering. Kvantovaya Elektronika (Moscow). 1992;19:688-690

33. Losev VF, Panchenko YN. Compression of a XeCl laser-pulse using stimulated brillouin-scattering. Kvantovaya Elektronika (Moscow). 1994;21:55-56

34. Baksht EH, Losev VF, Panchenko AN, Panchenko YN, Tarasenko VF. XeCl master oscillator with 300 ns pulse duration. Proceedings of SPIE. 2002;4747:88-92. DOI: 10.1117/12.460106

35. Baksht EH, Panchenko AN, Tarasenko VF. Discharge lasers pumped by generators with inductive energy storage. IEEE Journal of Quantum Electronics. 1999;35:261-266. DOI: 10.1109/3.748826

36. Rukin SN. High-power nanosecond pulse generators based on semiconductor opening switches (review). Instruments and Experimental Techniques. 1999;42:439-467

37. Konovalov IN, Losev VF, Liu J, Panchenko YN. Wide-aperture electric-discharge XeCl lasers. Proceeding of SPIE. 2004;5483:55-59. DOI: 10.1117/12.562962

38. Konovalov IN, Losev VF, Panchenko YN, Ivanov NG, Sykhov MY. Electric-discharge XeCl laser emitting 10-J, 300-ns pulses. Quantum Electronics. 2005;35:237-240. DOI: 10.1070/QE2005v035n03ABEH 0034 28

39. Ozaki T, Kieffer JC, Toth R, Fourmaux S, Bandule H. Experimental prospects at the Canadian advanced laser light source facility. Laser Particle Beams. 2006;24:101-106. DOI: 10.1017/S0263034606060150

40. Hofmann T, Sharp TE, Dane CB, Wisoff PJ, Wilson WL, Tittel FK, Szabo G. Characterization of an ultrahigh peak power XeF(C-A) excimer laser system. IEEE Journal of Quantum Electronics. 1992;28:1366-1375. DOI: 10.1109/3.135279

CHAPTER-2

CONTINUOUSLY TUNABLE TERAHERTZ-WAVE GENERATION WITH HIGH BRIGHTNESS

1. INTRODUCTION: AN OVERVIEW

Terahertz waves (wavelength, 30–3000 μm; frequency, 10–0.1 THz) are significant not only in the fields of basic science, such as molecular spectroscopy, molecular optics, plasma measurement, charged particle acceleration, and radio astronomy, but also in a variety of applications, including broadband wireless communication, nondestructive inspection, high-precision radar, and global environmental measurement. This is due to their higher directivity than infrared and their higher transmittances in the atmosphere and soft materials, such as microwaves, than infrared. Consequently, it is necessary to develop terahertz-wave sources that are continuously tunable, have a narrow linewidth (high-brightness), and have a high peak power in order to be extensively applicable in these applications. The terahertz (THz) region is comparatively unexplored due to the absence of commercially available high-brightness and continuously tunable sources, high-sensitive and fast detectors, and optics, which has led to the development of the frequency gap [1, 2, 3]. The field of science and engineering has experienced significant growth in the past two decades, transforming into a cross-disciplinary, international, and vibrant research activity [4].

Wavelength (frequency) conversion in nonlinear optical materials is a highly effective method for producing continuously tunable terahertz waves with high luminosity. This is due to the high conversion efficiency, bandwidth, wide tunability, and use at room temperature. The terahertz-wave parametric oscillator [7] was developed in the mid-1990s, following the initial proposal of a terahertz-wave source that utilized parametric wavelength conversion and was based on lithium niobate ($LiNbO_3$) crystals in the 1960s [5, 6]. At that time, the utmost peak output power of a terahertz wave was observed to be several milliwatts, while the tuning range was 1.1 to 1.6 THz (270–184 μm). The peak output power exceeded 55 kW [8, 9, 10, 11] and the tuning range expanded from 0.39 to 5.0 THz (750–60 μm) by using the current injection-seeded terahertz-wave parametric generator (is-TPG). This represents a 10 times and seven orders of magnitude increase, as illustrated in Figure 1. In Table 1, the characteristics of three typical intense terahertz-wave sources are presented: our injection-seeded terahertz-wave parametric generator (is-TPG), well-known intense terahertz-wave sources, a far-infrared free-electron laser (FIR-FEL) [12], and terahertz-wave pulse generation through optical rectification using a tilted-pulsefront excitation (OR) [13]. The is-TPG is a source of exceptional brightness in the terahertz region, exhibiting a broad tuning range. In this chapter, we delineate the process by which high-brightness terahertz waves are produced by the acoustic phonons of a nonlinear lithium niobate ($LiNbO_3$) crystal.

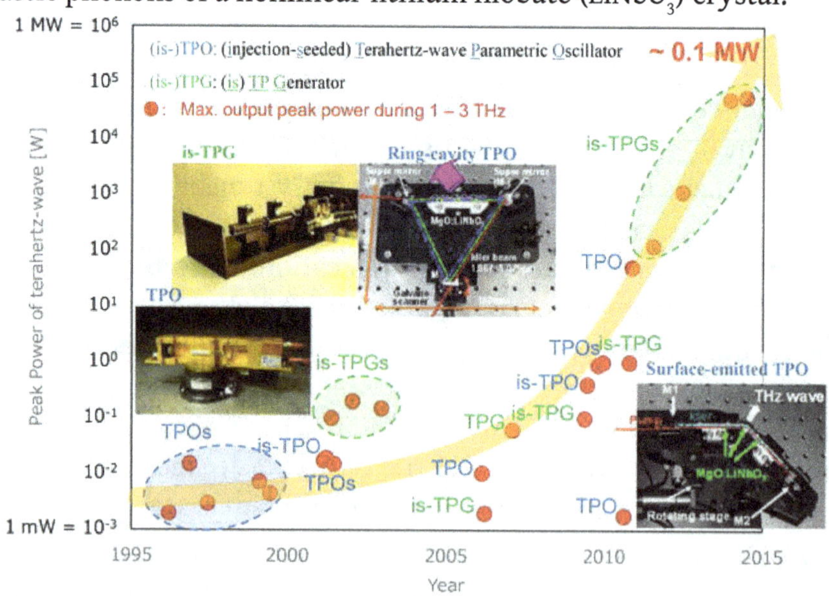

Figure 1. Development of parametric sources using $LiNbO_3$ in our group. Points represent the peak output power during 1–3 THz in each paper.

Table 1. **Characteristics of typical intense terahertz-wave sources: our injection-seeded terahertz-wave parametric generator (is-TPG), terahertz-wave pulse generation through optical rectification using a tilted-pulse-front excitation (OR), and a narrowband free-electron laser that works in the far-infrared region (FIR-FEL). The brightness temperature (T_b) is calculated as** $k_B T_B$ = *Peak power/[$(M^2)^2$ × linewidth]*. **OR generates broadband terahertz waves.**

	Peak power [kW]	Linewidth [GHz]	Tb [K]	Turning range
is-TPG	~ 100	~ 4	~ 10^{18}	0.4 – 5
OR	~ 5000	~ 1500	~ 10^{17}	Broadband
FIR-FEL	~ 2	~ 10	~ 10^{16}	1 – 5

2. TERAHERTZ-WAVE PARAMETRIC GENERATION

When the intense laser beams pass through a nonlinear optical crystal, the transverse photon and phonon wave fields become coupled and behave as new mixed photon-phonon states called polaritons. Broadband terahertz-wave generation results from efficient parametric scattering of laser light via polaritons [5, 6]. The polaritons exhibit phonon-like behavior in the resonant frequency region (near the transverse optical (TO)-phonon frequency ω_{TO}); however, they behave like photons in the nonresonant low-frequency region, as shown in Figure 2. Generation of narrowband terahertz waves can be achieved by applying an optical resonator (in the case of the terahertz-wave parametric oscillator (TPO)) or injecting a "seed" (in the case of the injection-seeded terahertz-wave parametric generator (is-TPG)) for the idler wave [14]. The wide tunability is accomplished simply by changing the angle between the incident pumping beam and the resonator axis (in the case of TPOs) or the wavelength and axis of the seeding beam satisfying the phase-matching condition (in the case of is-TPGs). In the parametric wavelength conversion process, a terahertz-wave signal photon and a near-infrared idler photon are created parametrically from a near-infrared pumping photon, according to the energy conservation law $\omega_p = \omega_T + \omega_i$ (where ω indicates frequency and p, T, and i denote the pumping, terahertz, and idler photons, respectively) and the momentum conservation law $\mathbf{k}_p = \mathbf{k}_i + \mathbf{k}_T$ (noncollinear phase-matching condition). This condition leads to the angle-dispersive characteristics of the idler and terahertz waves.

Thus, broadband terahertz waves can be generated depending on the phase-matching angle.

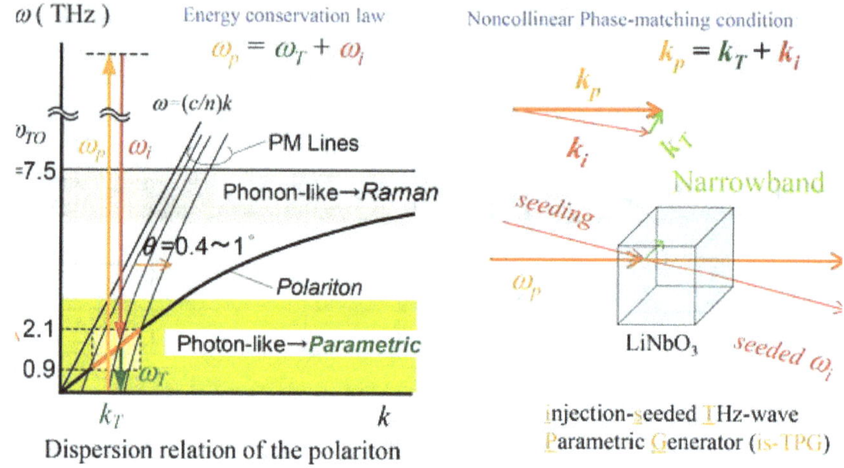

Figure 2. (Left) Dispersion relation of the polariton. (Right) Noncollinear phase-matching condition.

In this experiment, we used a magnesium oxide (MgO)-doped $LiNbO_3$ crystal. The large figure of merit (FOM = $4d_{eff}^2/n_{NIR}^2 n_{THz} \alpha_{THz}^2 \sim 10$, d_{eff}; the effective nonlinear coefficient, n_{NIR} and n_{THz}; the refraction indices in the near infrared and terahertz range, α_{THz}; the absorption coefficient for the terahertz wave) [13] of $LiNbO_3$ at room temperature makes this well-known nonlinear crystal ideal for such an application. The gain curve of the terahertz-wave parametric generation is determined by the parametric gain and absorption coefficients in the terahertz region. Figure 3 shows the pumping intensity dependence (0.1, 0.2, 0.4, 0.8, 1.6, and 3.2 GW/cm^2) of calculated gain curves [15]. As the pumping intensity increases, the gain coefficient also increases in whole frequency region, and the maximum value in the gain curve moves toward higher frequencies. All gain curves have a broad bandwidth, with a drop appearing at around 2.6 THz. This is because the low-frequency modes of doped MgO in the $LiNbO_3$ work as crystal lattice defects. Under the noncollinear phase-matching condition, the effective gain curve depends on both the intensity and the beam diameter of the pumping beam. Figure 4 shows the pumping beam diameter dependence of calculated effective gain curves. When an Nd:YAG laser (λ = 1064 nm) is used to generate the pumping beam and an MgO:$LiNbO_3$ crystal is used as the nonlinear optical crystal, the effective gain coefficient is given [16, 17, 18] by where $\alpha_{crystal}$ is the absorption coefficient of the terahertz wave in the MgO:$LiNbO_3$ crystal, φ is the phase-matching angle between the pumping beam and the terahertz wave, g_0 is the parametric gain under the low-loss limit, and r_p is the diameter of the pumping beam. When the pumping beam intensity is 1 GW/cm^2, all gain curves (pumping beam diameter 0.5, 1, 2, and 5 mm) have enough gain coefficient over a broad range extending from less than 1 THz to more than 3 THz. As the pumping beam diameter becomes larger, the gain coefficient also increases, and the maximum value of the gain curve moves toward lower frequencies. As a result, we can optimize the tuning curves by controlling these parameters.

$$g_T = \alpha\left[\left(1 + 16 \cos \varphi \left(g_0/\alpha\right)^2\right)^{1/2} - 1\right]/2,$$

$$\alpha = \alpha_{crystal} + \alpha_{beam},$$

$$\alpha_{beam} \propto \sin \varphi / r_p,$$

When a high-intensity laser beam propagates through a nonlinear crystal, a number of nonlinear processes occur, such as the following: second- or higher-harmonic generation (SHG or HHG); difference- or sum-frequency generation (DFG or SFG); optical parametric generation, amplification, or oscillation (OPG, OPA, or OPO); stimulated Raman or Brillouin scattering (SRS or SBS); four-wave mixing; optical rectification (OR); multiphoton absorption; and the Kerr and Pockels effects. Of these, we revealed that the parametric wavelength conversion near the lattice resonance induced by SRS is significantly inhibited by SBS; however, this nonlinear process has long been ignored. In the previous research done by authors, the conversion efficiency in energy from an infrared pumping beam to a terahertz wave was less than 10^{-7}. It has long been thought that this is the limit of the conversion efficiency using parametric wavelength conversion using $LiNbO_3$ pumped by nanoseconds Nd:YAG lasers (duration, 10–25 ns) [14]. However, when a photon of the pumping beam (1064 nm) creates two photons (idler beam and terahertz wave (100–1000 μm), in principle, the conversion efficiency reaches 10^{-2}–10^{-3} according to the Manley-Rowe relations because the wavelength of the terahertz wave is about 10^2–10^3 times longer wavelength than that of the pumping beam. In our experiment, an infrared pumping beam excites acoustic phonons in $LiNbO_3$, and SRS of the pumping beam generates terahertz waves and an idler beams. We calculated both gain coefficient of the SRS and the SBS in the previous condition; the gain coefficient of SBS has 1000 times larger gain than that of the SRS [19, 20, 21, 22, 23, 24]. Typically, the SBS gain reaches the steady state within 10 lifetimes of the acoustic phonon of crystal [25], within about 1.5 ns in $LiNbO_3$ [24].

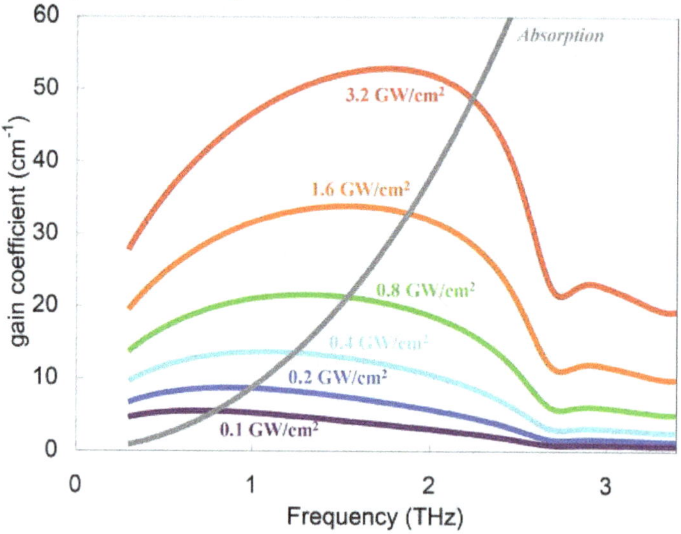

Figure 3. Pumping intensity dependence of calculated gain coefficient using MgO:$LiNbO_3$ pumped by Nd:YAG laser, when the pumping intensities were 0.1, 0.2, 0.4, 0.8, 1.6, and 3.2 GW/cm^2.

For efficient wavelength conversion, the pulse width of the pumping beam should be enough less than this, but the pulse width limits the linewidth of the generated terahertz waves. By applying a single-mode oscillated microchip Nd:YAG laser [26] with a sub-nanosecond (several hundreds of picoseconds) "pulse gap" pulse width [27] as a pumping source, a high-efficiency and narrow-linewidth wavelength conversion can be performed by the SRS without the SBS. Additionally, when the intensity of the pumping beam is too high, second-order stoke (idler) beams can be generated, which do not contribute to the generation of terahertz waves as they undergo strong absorption. We thus precisely controlled both pumping and seeding intensity and diameter as well as the nonlinear crystal length.

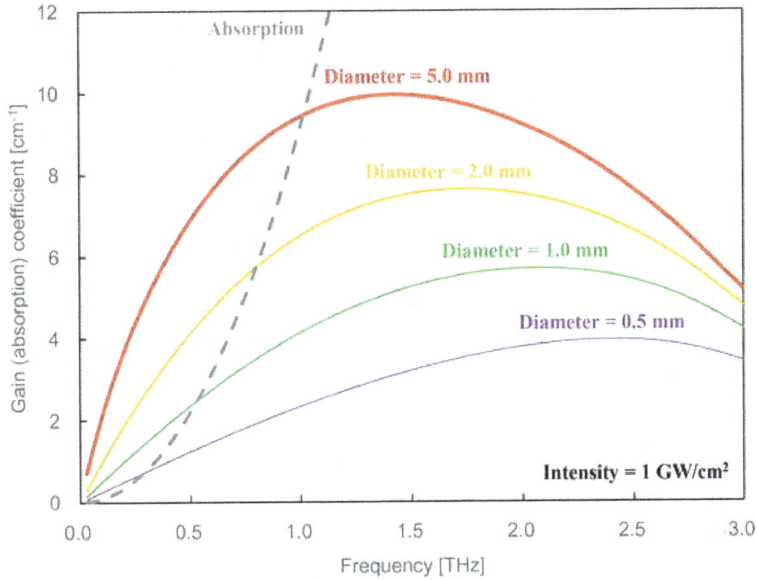

Figure 4. Pumping diameter dependence of effective calculated gain coefficient, when the pumping beam diameters were 0.5, 1, 2, and 5 mm. The pumping intensity was 1 GW/cm^2.

3. EXPERIMENT

The experimental apparatus, shown in Figure 5, consists of a pumping source (pulsed, Nd:YAG laser), a seeding source (CW, external cavity diode laser (ECDL)), amplifiers (for both pumping and seeding beams), and the nonlinear crystal (MgO:$LiNbO_3$). The pumping source is a diode end-pumped microchip Nd^{3+}:YAG laser passively Q-switched by Cr^{4+}:YAG saturable absorber. This configuration enables the low-order axial and transverse mode laser oscillation, whose linewidth is below 0.009 nm. The laser delivers more than 1.4 MW peak power pulses (energy/pulse, > 0.6 mJ/pulse; duration, ~ 420 ps) at 100 Hz repetition rate with a M^2 factor of less than 1.1 [26]. The pumping beam is amplified by two amplifiers in double-pass configurations. Each amplifier has 0.7 at.% doped Nd^{3+}:YAG with a length and diameter of 70 and 3 mm, transversely pumped by 200 W laser diodes (wavelength, 808 nm) in a threefold geometry. Amplified beam is extracted by a polarization beam splitter (PBS). The seeding beam from an ECDL is also amplified by an **Yb**-doped fiber amplifier. Owing to the grating and confocal arrangement, the noncollinear phase-matching condition is satisfied automatically depending on the wavelength of the seeding beam [28]. The diameter of both beams is the same on the crystal input surface. We used a 50-mm-long nonlinear $MgO:LiNbO_3$ crystal with antireflection coating for a pumping beam. A prism made by high-resistivity silicon placed on the output surface of the nonlinear crystal works as an efficient output coupler only for the terahertz waves to avoid the total internal reflection of the terahertz waves on the crystal output side surface. For an optimization of terahertz-wave emission, the pumping region within the nonlinear crystal must be as close as possible to the output surface, because of the large absorption coefficient of the $MgO:LiNbO_3$ crystal in the terahertz range (10–100 cm^{-1}). The distance between the output surface and the beam center was precisely adjusted to obtain a maximum terahertz-wave output, and it was approximately equal to the pumping beam radius. The terahertz-wave output extracted through the Si-prism coupler was collimated, focused, attenuated, modulated, and then measured using a calibrated pyroelectric detector covered by thick black polyethylene sheet. The temporal waveform and linewidth of the terahertz wave were measured by a Schottky barrier diode (SBD) and a pair of scanning metal mesh plates.

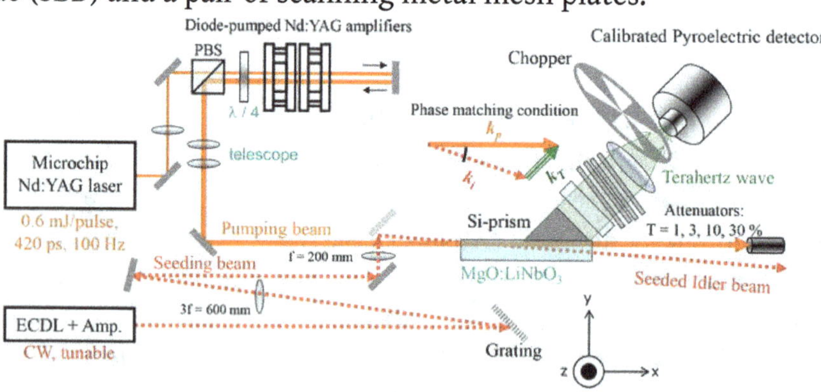

Figure 5. Experimental apparatus for an is-TPG.

4. RESULT AND DISCUSSION

Figure 6 shows the tuning curves of two is-TPGs fabricated using our design obtained by scanning the wavelength of seeding beam. When the pumping beam diameter and energy are 1.5 mm and 20 mJ/pulse, and the seeding beam power is 800 mW (continuous wave), the tunable range of the terahertz wave is 0.7–3 THz (430–100 μm). The maximum output peak power is more than 55 kW (BT ~ 10^{18} K, brightness ~ 0.2 GW/

sr·cm^2) at around 1.8 THz. This source has a broad tuning range, with a flat region around 1.6–2.6 THz. The terahertz-wave output decreased in the low- and high-frequency regions (below 1.6 and above 2.6 THz) because of a low parametric gain and high absorption coefficient [29] in these regions, respectively. From the is-TPG, the pulsed terahertz waves are generated by 100 Hz (by 10 ms); however, the pyroelectric detector we used in the experiment only gives an average power. We therefore used an optical chopper to measure the average power. We estimated the energy/pulse from the calibrated average power, and the maximum energy/pulse was more than 5.5 μJ/pulse. When the thick glass plate as an IR pass filter was inserted, the output signal from the detector completely disappeared. We measured the duration of generated terahertz wave around 100 ps by the SBD, corresponding to peak power of more than 55 kW at 1.8 THz. The conversion efficiency in energy from pumping beam to terahertz wave is about 10^{-4} in this case. The tuning curve for a 2.2-mm-diameter pumping and seeding beam is also shown in Figure 6. In that case, the tunable range is 0.39–2 THz (760–150 μm). The tuning curve has an extremely broad bandwidth, and the lowest frequency (longest wavelength) of 0.39 THz (760 μm) was also the lowest frequency (longest wavelength) achieved in our experiment. The maximum terahertz-wave output is more than 7 kW (energy, 0.7 μJ/pulse; duration, 100 ps), which occurs near 330 μm (0.9 THz) when the pumping energy is 15 mJ/pulse and seeding power is 2400 mW. The conversion efficiency is about 10^{-5} in this case.

These tuning curves depend on the gain curves, respectively. In the case of large-diameter pumping and seeding beams, the tuning curve has been shifted toward lower frequency (longer wavelength); the output peak power in the sub-terahertz range has been increased. These are because the parametric gain in the low-frequency (long-wavelength) region was increased by expanding the beam diameters as shown in Figure 4; that is, wavelength conversion in the region was effectively achieved. Meanwhile, in the high-frequency (short-wavelength) region, both tuning range and output peak power have decreased, because of the large absorption associated with increasing propagation distance in the crystal.

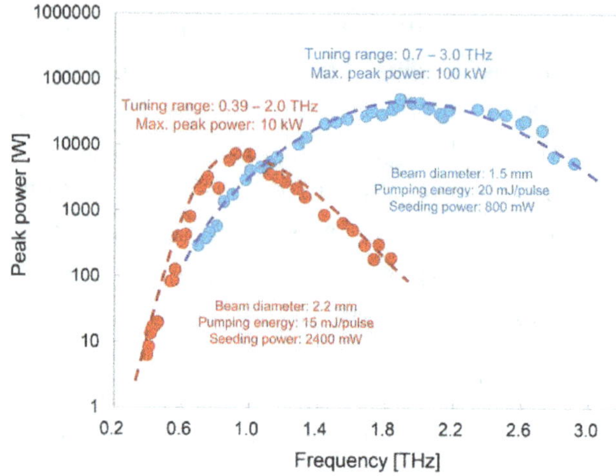

Figure 6. The tuning curves of two is-TPGs with pumping beam diameters of 1.5 mm, represented by blue solid line and 2.2 mm by red one.

Figure 7 shows a beam profile of terahertz wave from the is-TPG measured by an imager for terahertz wave. The generated terahertz wave is collimated and focused by the Tsurupica cylindrical (f = 100 mm) and aspherical (f = 50 mm) lens. When the wavelength of the terahertz wave is 200 μm, the spot size is less than 220 μm (full width at half maximum). We estimated the M^2 value less than 1.1, and the brightness was $B = Pp/(\lambda M^2)^2$ ~ 0.2 GW/sr·cm². The intensity and electric field were 0.3 GW/cm² and 0.5 MV/cm at around 2.0 THz, respectively, at the focused point.

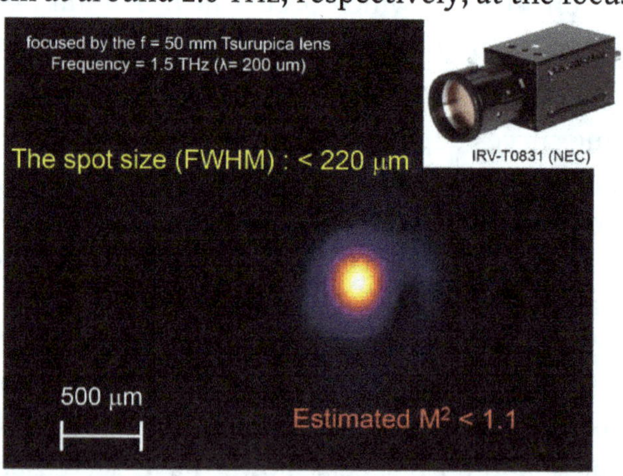

Figure 7. A beam profile of a terahertz wave from the is-TPG measured by a terahertz-wave imager (IRV-T0831, NEC). The terahertz wave is focused by the f = 50 mm lens. The spot size is less than 220 mm (FWHM) at 1.5 THz.

Figure 8 presents an example of wavelength and linewidth measurement by a scanning Fabry-Perot etalon consisting of two metal mesh plates. The horizontal axis represents the distance between metal meshes, and the vertical axis represents the energy of the transmitted terahertz wave. The metal meshes were made of nickel and had periods of 45 μm and reflectance of about 98% at 0.75 THz. As the distance between metal mesh plates increases, intensity peaks are observed periodically. In this case, the estimated wavelength (frequency) of the terahertz wave is about 398 μm (0.754 THz), and the linewidth is less than 5 GHz. This linewidth is a near the Fourier transform limit for the terahertz-wave pulse with a sub-nanosecond duration.

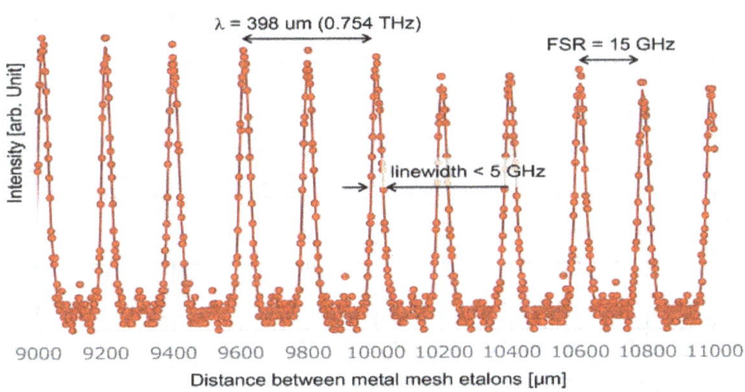

Figure 8. An example of the wavelength and linewidth measurements. In this case, the wavelength (frequency) of the terahertz wave is approximately 398 μm (0.754 THz), and the linewidth is less than 5 GHz. This linewidth is near the Fourier transform limit in the sub-nanosecond pulse.

5. CONCISE SUMMARY

In this paper, we have demonstrated the continuous tunable generation of terahertz waves with a high peak power, narrow linewidth, and in a MgO:LiNbO3 crystal through wavelength conversion. These are the outcomes of the suppression of the SBS in a nonlinear crystal through the use of a sub-nanosecond pumping pulse. The power calibration of terahertz-wave detectors is significantly influenced by the high-brightness and continuously tunable source. In general, the power calibration is based on calorimetry as a traceable international standard; however, there is no power standard in the terahertz region. In this instance, the power levels derived from two types of pre-calibrated detectors, a calorimetric device and a pyroelectric device, using the same terahertz beam, were comparable [30]. This was unexpectedly readily perceptible through touch; the terahertz wave was perceived as comparable to a 100 Hz (rep. rate) stimulation. The conversion efficiency that is observed under our experimental conditions is approximately 10-4. This is due to the fact that the terahertz wave that is generated within the crystal is absorbed by the nonlinear crystal while it is propagating to the crystal surface and is influenced by Fresnel loss on the boundary surfaces. Moreover, the crystal's parametric gain (absorption) of the terahertz wave in $LiNbO_3$ could be either increased or diminished by cooling [31]. At liquid nitrogen temperatures, the conversion efficiency is enhanced by a factor of at least 10. In this instance, the anticipated brightness, brightness temperature, peak power, and electric field of the terahertz wave are all greater than 4 GW/sr·cm^2, 1019 K, 1 MW, and 2 MV/cm, respectively, when the pumping energy is 50 mJ/pulse. These values are obtained from our continuously tunable is-TPG, which operates in the narrowband domain. Certain applications necessitate high-brightness terahertz waves, such as the observation of two- or multiphoton absorption to specific excitation states [32, 33]. Novel applications in the field of terahertz nonlinear optics will be facilitated by the generation of ultra-high-brightness quasi-monochromatic terahertz-wave pulses (several hundreds of cycles) with field levels in the megavolt per centimeter (~ MV/cm) range. These pulses have a narrow (~ GHz) linewidth and a peak power of megawatts (~ MW). In addition, we illustrated how to optimize the tuning curve of the is-TPG by regulating the interaction volume and pumping intensity.

In the future, it is imperative that we strive to produce a beam with a higher luminosity and a broader tuning range for the purpose of conducting applied research. As a method of facilitating nonthermal free target energy-level control and measurement, extreme high-brightness terahertz-wave generation has garnered attention in recent years. The implementation of a terahertz-wave control and measurement system would enable the development of novel applications in the terahertz region and the potential resolution of a variety of contemporary societal challenges. Not only could this system be a powerful tool for solving real-world problems, but it could also be used to manipulate or alter atoms, molecules, chemical materials, proteins, cells, chemical reactions, and biological processes, as well as for real-time spectroscopic measurement/imaging and 3D fabrication. We anticipate that these methods will introduce novel applications and new domains.

AT A GLANCE

An intriguing frequency range is found in the "frequency gap" zone between millimeter wave and infrared, known as the terahertz (THz) wave. Despite the development of novel techniques for producing terahertz radiation, the majority of sources are unable to generate terahertz waves that possess both high brightness (characterized by high peak power and narrow linewidth) and continuous tunability. In this study, we present a method for producing terahertz waves with great brightness and the ability to be constantly adjusted in frequency. This is achieved using parametric wavelength conversion in a nonlinear crystal. The resulting terahertz waves are even brighter than those generated by specialized sources like far-infrared free-electron lasers. We have demonstrated a new method of converting wavelengths utilizing stimulated Raman scattering in lithium niobate, a nonlinear crystal. This method does not need stimulated Brillouin scattering and utilizes a recently designed microchip Nd:YAG laser. In addition, we demonstrate the method of enhancing the tuning curve by manipulating the pumping and seeding beam. These sources show great potential for expanding applied research in the terahertz range. We anticipate that this source will facilitate the exploration of new research areas, such as nonlinear optics in the terahertz region.

REFERENCES

1. Tonouchi M. Cutting-edge terahertz technology. Nature Photonics. 2007;1:97-105

2. Chamberlain JM. Where optics meets electronics: Recent progress in decreasing the terahertz gap. Philosophical Transactions of the Royal Society of London/A. 2004;362:199-213

3. Sherwin M. Terahertz power. Nature. 2002;420:131-132
4. Linfield E. Terahertz applications–A source of fresh hope. Nature Photonics. 2007;1:257-258
5. Henry CH, Garrett CGB. Theory of a parametric gain near a lattice resonance. Physics Review. 1968;171:1058-1064
6. Piestrup MA, Fleming RN, Pantell RH. Continuously tunable submillimeter wave source. Applied Physics Letters. 1975;26(8):418-421
7. Kawase K, Sato M, Taniuchi T, Ito H. Coherent tunable THz wave generation from $LiNbO_3$ with monolithic grating coupler. Applied Physics Letters. 1996;68(18):2483-2485
8. Minamide H et al. Kilowatt-peak terahertz-wave generation and sub-femtojoule terahertz-wave pulse detection based on nonlinear optical wavelength-conversion at room temperature. Journal of Infrared, Millimeter, and Terahertz Waves. 2014;35(1):25-37
9. Hayashi S'i, Nawata K, Taira T, Shikata J, Kawase K, Minamide H. Ultrabright continuously tunable terahertz-wave generation at room temperature. Science Reports. 2014;4. Art. No. 5045
10. Hayashi S'i, Nawata K, Takida Y, Tokizane Y, Kawase K, Minamide H. High-brightness continuously tunable narrowband subterahertz wave generation. IEEE Transactions on Terahertz Science and Technology. 2016;6(6):858-861
11. Murate K, Hayashi S'i, Kawase K. Multi-wavelength terahertz-wave parametric generator for one-pulse spectroscopy. Applied Physics Express. 2016;9:082401
12. Rarnian G. The new UCSB free-electron lasers. Nuclear Instruments and Methods in Physics Research, Section A. 1992;318:225-229
13. Hebling J, Yhe KL, Hoffmann MC, Bartal B, Nelson KA. Generation of high-power terahertz pulses by tilted-pulse-front excitation and their application possibilities. Journal of the Optical Society of America B: Optical Physics. 2008;25:B6-B19
14. Kawase K, Shikata J, Imai K, Ito H. Transform-limited, narrow-linewidth, terahertz-wave parametric generator. Applied Physics Letters. 2001;78(19):2819-2821

15. Shikata J, Kawase K, Taniuchi T, Ito H. Fourier transform spectrometer with a terahertz-wave parametric generator. Japanese Journal of Applied Physics. 2002;41(Part 1, No. 1):134-138

16. Schwarz UT, Maier M. Frequency dependence of phonon-polariton damping in lithium niobate. Physical Review. B, Condensed Matter. Mar. 1996;53(9):5074-5077

17. Schwarz UT, Maier M. Damping mechanisms of phonon polaritons, exploited by stimulated Raman gain measurements. Physical Review. B, Condensed Matter. Jul. 1998;58(2):766-775

18. Takida Y et al. THz-wave parametric gain of stimulated polariton scattering. Physical Review A: General Physics. Apr. 2016;93(4):043836-1-043836-4

19. Agramal GP. Stimulated Brillouin scattering. In: Kelly PL, Kaminow I, Agrawal GP, editors. Nonlinear Fiber Optics. 3rd ed. San Diego: Academic Press; 1995. pp. 355-362

20. Weis RS, Gaylord K. Lithium niobate: Summary of physical properties and crystal structure. Applied Physics A: Materials Science & Processing. 1985;37:191-203

21. Dmitriev VG, Gurzadyan GG, Nikogosyan DN. Properties of nonlinear optical crystals. In: Schawlow AL, Tamir T, Siegman AE, editors. Handbook of Nonlinear Optical Crystals. 3rd Revised Edition. Berlin: Springer; 1990. p. 119

22. Hinkov V, Barth M, Dransfeld K. Acoustic properties of proton exchanged $LiNbO_3$ investigated by Brillouin scattering. Applied Physics A: Materials Science & Processing. 1985;38:269-273

23. Sussner H, Vacher R. High-precision measurements of Brillouin scattering frequencies. Applied Optics. 1979;18:3815-3818

24. de Bemabe A, Prieto C, de Andres A. Effect of stoichiometry on the dynamic mechanical properties of $LiNbO_3$. Journal of Applied Physics. 1996;79:143-148

25. Faris GW, Jusinski LE, Hickman AP. High-resolution stimulated Brillouin gain spectroscopy in glasses and crystals. Journal of the Optical Society of America B: Optical Physics. 1993;10:587-599

26. Sakai H, Kan H, Taira T. 1 MW peak power single-mode high-brightness passively Q-switched Nd^{3+}:YAG microchip laser. Optics Express. 2008;16:19891-19899

27. Taira T. Domain-controlled laser ceramics toward giant micro-photonics. Optical Materials Express. 2011;1:1040-1050

28. Imai K, Kawase K, Minamide H, Ito H. Achromatically injection-seeded terahertz-wave parametric generator. Optics Letters. 2002;27:2173-2175

29. Hayashi S et al. Output power enhancement of a palmtop terahertz-wave parametric generator. Applied Optics. 2007;46:117-123

30. Dobroiu A, Otani C. Calibration of terahertz-wave detectors: comparison procedure and error estimation. Proceedings of the 3rd EOS Topical Meeting on Terahertz Science & Technology; 2012. ISBN: 978-3-9815022-1-3

31. Shikata J, Sato M, Taniuchi T, Ito H, Kawase K. Enhancement of terahertz wave output from $LiNbO_3$ optical parametric oscillator by cryogenic cooling. Optics Letters. 1999;24:202-204

32. Kulipanov GN et al. Research highlights from the Novosibirsk 400 W average power THz FEL. Terahertz Science and Technology. 2008;1:107-125

33. Yokoyama K, Matsuoka L, Kasajima T, Tsubouchi M, Yokoyama A. Quantum control of molecular vibration and rotation toward the isotope separation. Proceedings of the 5th Asian Symposium on Intense Laser Science (ASILS-5). 2010:113-119

CHAPTER-3

HYBRID FEMTOSECOND LASER SYSTEMS WITH HIGH POWER AND HIGH INTENSITY CONTRAST

1. INTRODUCTION: AN OVERVIEW

The development of high-power femtosecond laser systems was achieved through the utilization of the chirped pulse amplification (CPA) approach [1]. In recent years, PW-class Ti:sapphire laser systems have been successfully demonstrated on a global scale [2, 3, 4, 5, 6].

In order to achieve a high peak pulse power, it is necessary to have a large amount of pulse energy delivered in a short duration. The laser fluence in a laser amplifier system is limited by the potential dangers of damage. The maximum energy of the amplified laser pulse is constrained by the dimensions of the present optical elements, such as Ti:sapphire crystals and diffraction gratings found in optical compressors. In order to achieve a high peak pulse power in CPA systems using existing optical components, it is crucial to provide the laser energy in laser pulses with durations in the region of 10 femtoseconds. The duration of the recompressed amplified pulse is inversely related to the spectral bandwidth, which includes the phase-locked spectral components, indicating a flat-phase spectral bandwidth. The narrowing of the spectral band and the elongation of the duration of the recompressed amplified pulse occur due to gain narrowing and redshifting in Ti:sapphire crystals [7]. Special approaches such as optical cross-polarized wave (XPW) production and spectral filters can be used to preserve a wide range of amplified laser pulses in CPA laser systems. The use of acousto-optic programmable dispersion filters (AOPDFs) allows for the correction of high-order phase distortions, resulting in a flat spectral phase throughout a wide bandwidth [8-9]. In various research applications, a concentrated laser beam with a very high intensity is necessary. The capacity to precisely concentrate the laser beam into a minute area is a crucial characteristic of high-power CPA laser systems. The degree of tight focusing is heavily influenced by the quality of the amplified pulse beam wavefront. Thermal loading of Ti:sapphire crystals is a primary cause of wavefront aberrations in CPA systems. The Strehl ratio (SR) quantifies the focused beam intensity in relation to an undisturbed flat wavefront with the same intensity profile as the real beam [10]. By concentrating 300 terawatt femtosecond laser beams into spots with a diameter of a few micrometers, a peak intensity of 2 × 10^22 watts per square centimeter has been achieved [11]. Tightly concentrating 10-PW femtosecond laser pulses is anticipated to yield a peak power exceeding 1023 W/cm2.

If a laser intensity of about 10^{11} W/cm^2 is attained on the target before the main laser pulse, the generated pre-plasma could disturb the experiment. High-intensity contrast becomes a crucial laser beam parameter for accessing high-field physics in various experimental targets. Some techniques for improving the intensity contrast of laser emission, such as saturable absorbers [12, 13] and XPW [14, 15, 16], were used inside high-power femtosecond CPA laser systems. Plasma mirrors, based on self-induced plasma shuttering, were proposed for improving intensity contrast after the temporal compression of amplified chirped laser pulses [17, 18]. Reaching an intensity contrast in the range of 10^{12} represents a challenging task for a multi-PW all Ti:sapphire CPA laser. Optical parametric chirped pulse amplification (OPCPA) in nonlinear crystals provides large amplification spectral bandwidth and improves the intensity contrast of the amplified pulses outside the temporal window of the parametric amplification process [4, 19]. In case of OPCPA with high-energy laser pulses, an important technical problem consists in the generation of a single pump beam with simultaneously difficult-to-accomplish specifications: hundreds of Joules laser pulse energy, about one nanosecond pulse duration, spatial and temporal smooth and nearly flat intensity profile, very good stability from pulse to pulse, high repetition rate.

Hybrid femtosecond lasers combine OPCPA in nonlinear crystals at low-medium energy with CPA in large size Ti:sapphire crystals at high energy. A key feature of high-power 10-fs laser systems consists in the adaptation of the parametric amplification phase matching bandwidth of nonlinear crystals to the spectral gain bandwidth of laser amplifying Ti:sapphire crystals. The spectrum bandwidth for phase-matching near a wavelength of 800 nm in beta-barium borate (BBO) crystals, when pumped by green lasers, is nearly identical to the gain bandwidth of Ti:sapphire crystals. In this scenario, it is possible to maintain a wide spectral gain bandwidth over the entire hybrid amplification chain. It is possible to obtain femtosecond pulses that are recompressed and have both high power and great intensity contrast. The use of OPCPA in BBO crystals to achieve energy levels up to mJ in the Front-End, followed by CPA in large-aperture Ti:sapphire crystals to reach energy levels of 10/100 Joules, is an appropriate solution for femtosecond laser systems in the petawatt (PW) class.

The text describes a few globally advanced hybrid amplification high-power femtosecond laser systems. The hybrid amplification arrangement is deemed suitable for the 2 × 10 PW femtosecond laser system at the Extreme Light Infrastructure: Nuclear Physics (ELI-NP) facility.

2. CHIRPED PULSE AMPLIFICATION IN BROAD SPECTRAL BANDWIDTH LASER MEDIA

The principle of CPA in broad gain bandwidth laser media (e.g., Ti:sapphire crystals) is presented in Figure 1. Femtosecond laser pulses generated by a large spectral bandwidth oscillator are temporally stretched with dispersive optical elements, in most cases diffraction gratings, up to few-hundred picoseconds or about one nanosecond pulse duration.

The Ti:sapphire laser is a four-level system as depicted by a simplified energy level diagram in Figure 1. Ti:sapphire crystals are optically pumped by nanosecond green lasers. By absorption of pump laser photons, the Ti atoms are raised from the ground energy level E1 to the spectral band E4. The excited atoms are rapidly transferred by nonradiative transitions from the absorption band E4 to the upper laser energy level E3. The spontaneous fluorescence lifetime of Ti atoms on the upper laser level is about 3 μs. The Ti atoms are accumulated on the upper laser level giving rise to a population inversion between E2 and E3 laser levels. Under these conditions, an input laser pulse with photon energy quanta corresponding to the energy difference between the E3 upper level and the E2 lower level is amplified by stimulated laser transitions between E3 and E2 levels. The generated laser radiation is coherently added to the input radiation.

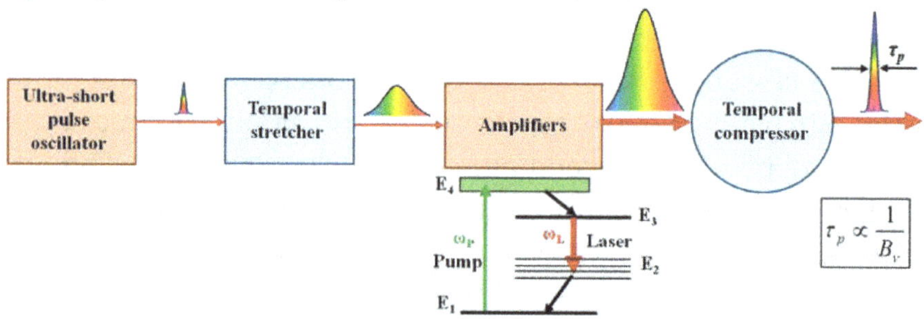

Figure 1. Chirped pulse amplification (CPA) in Ti:sapphire laser crystals. B_ν, amplified pulse frequency bandwidth; τ_p, temporally compressed pulse duration.

To get the population inversion between laser energy levels for laser amplification, we essentially need an efficient absorption of pump photons in Ti:sapphire crystals. The energy accumulated in the upper laser level can be the result of pumping with single or multiple pump laser beams. Angles between pump beams and seed pulse beam are noncritically defined and are practically imposed by the amplifier geometry.

Because the Ti atoms lifetime is in the range of few-μs, an acceptable delay between pump laser pulses and input stretched laser pulses is in the nanoseconds range. This temporal synchronization can be easily obtained with electronic devices.

Pulse duration of the recompressed pulse is inversely proportional to the optical frequency bandwidth which contains all phase-locked spectral components [20]. The highest amplification gain is obtained near the central wavelengths (790–800 nm) of the Ti:sapphire fluorescence spectrum, engendering the "gain narrowing" effect of the amplified laser pulse spectral band (Figure 2a). In the regenerative amplifiers and multi-pass amplifiers, with many passes through the laser amplifying media and high amplification factor, the effect of gain narrowing significantly contributes to the decrease of the spectral bandwidth of the amplified pulses (Figure 2b).

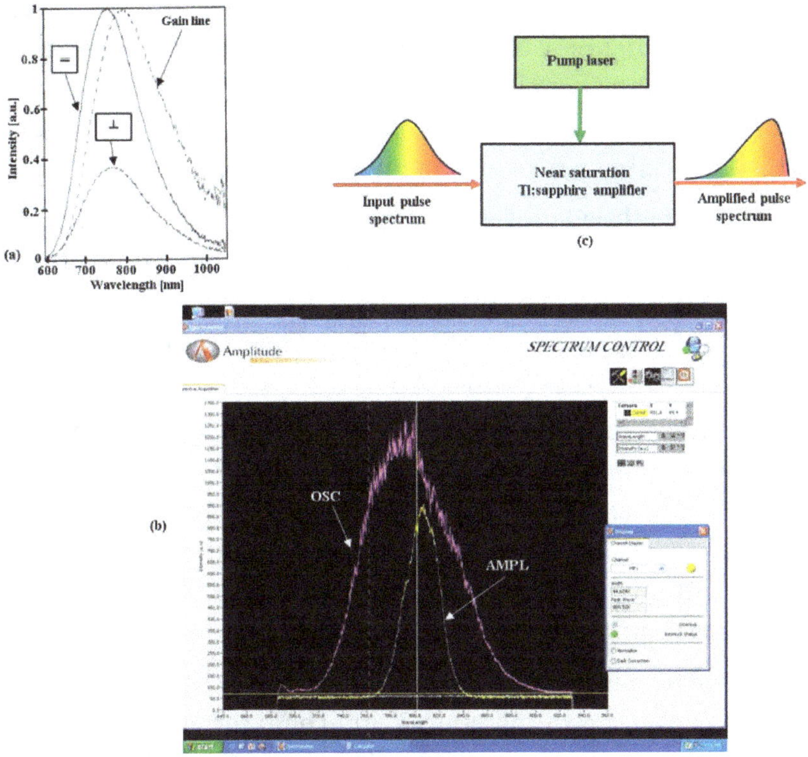

Figure 2. Gain narrowing and redshifting in Ti:sapphire amplifiers. (a) Polarized fluorescence spectra and calculated gain line for an optical c-axis normal cut Ti:sapphire rod; =, c-axis parallel polarization; ⊥, c-axis normal polarization. (b) Spectrum narrowing and redshifting after amplification in an all Ti:sapphire TW-class laser manufactured by Amplitude Technologies for the National Institute for Laser, Plasma, and Radiation Physics, Bucharest-Magurele; OSC: femtosecond oscillator spectrum; AMPL: spectrum after amplification. (c) Redshifting effect in nearly saturation operated Ti:sapphire amplifiers.

High-energy extraction efficiency can be obtained if laser amplifiers are working near the saturation regime, where the input laser pulse fluence is higher than the saturation fluence of the amplifying laser medium [20]. In this case, almost all accumulated energy on the upper level of the laser medium could be extracted and added to the input pulse energy [20]. The "red" spectral components travel in the leading edge of the temporally stretched pulse, whereas the "blue" spectral components are delayed in the trailing edge. In the amplifiers working near the saturation regime, due to the significant depletion of the upper laser-level population, the amplification factor of the "red" spectral components coming first in the amplifying medium is higher than that of the "blue" spectral components arriving on the trailing edge of the stretched pulse. The result is a redshift of the amplified laser pulse spectrum, associated with a spectrum narrowing (Figure 2b and c).

Stretched amplified pulses are recompressed in a temporal stretcher with diffraction gratings, where "red" spectral components are delayed compared to the "blue" components. Both "gain narrowing" and "redshifting" effects contribute to the increase of the amplified pulse duration after temporal recompression.

The amplified spontaneous emission (ASE), which takes place in the laser media as long as the population inversion between the upper and lower laser levels exists, deteriorates the picosecond intensity contrast of femtosecond laser systems. By all Ti:sapphire amplification, it is very difficult to attain more than 10^{11} intensity contrast of femtosecond pulses, as it is required in case of PW-class femtosecond laser systems.

Dissipated heat in the active medium is given by the energy difference between the absorbed pump energy and the laser emitted energy. The thermal loading of the Ti:sapphire crystals produces beam wavefront distortions and phase dispersions of the spectral components of the large bandwidth laser pulses. It results in a poor beam focusing and an increase of the recompressed pulse duration.

3. BROADBAND OPTICAL PARAMETRIC AMPLIFICATION

Optical parametric amplification (OPA) is practically an instantaneous process without laser energy accumulation in the amplifying medium.

By absorption of pump photons with ω_p frequency, the crystal molecules leave from their ground energy level E_1 to an excited intermediate higher energy level E_2 (Figure 3a). While an excited molecule returns to its initial ground state, a photon with ω_s signal frequency and one "idler" photon with $\omega_i = \omega_p - \omega_s$ are simultaneously created. This optical nonlinear process is very rapid compared to the signal and pump pulse duration.

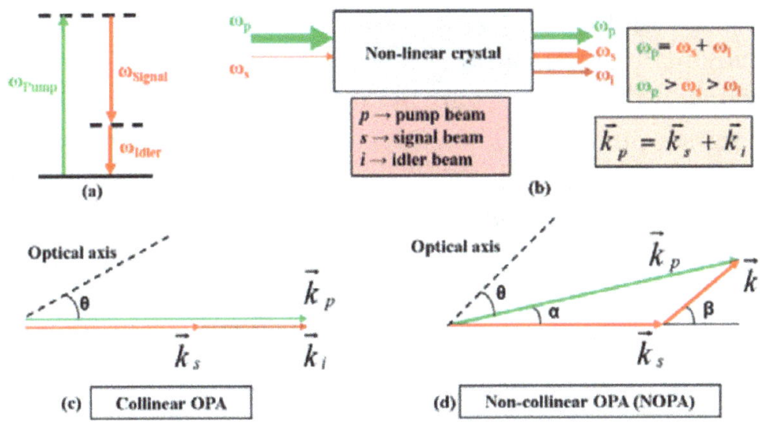

Figure 3. Optical parametric amplification in nonlinear crystals. (a) OPA energy level diagram. (b) Principle of OPA in a nonlinear crystal. (c) Collinear OPA geometry. (d) Noncollinear OPA (NOPA) geometry.

A fraction of the pump beam energy is transferred to the signal beam. At the output of the nonlinear crystal, we get an amplified signal beam, a new generated idler beam, and a residual pump beam (Figure 3b).

Amplification takes place only if the seed pulse and the pump pulse are spatially and temporally overlapped in the nonlinear crystal, in a collinear (Figure 3c) or a noncollinear geometry (Figure 3d). In case of nanosecond pulses OPA, temporal overlapping can be obtained by electronic synchronization of the pump pulsed laser with the signal pulses. In case of femtosecond/picosecond pulses, temporal overlapping can be obtained only by optical synchronization of the interacting laser pulses.

The parametric amplification is produced under conditions of photon energy conservation and wave-vector phase matching, only for a certain orientation of the crystal and for well-defined angles between the wave vectors of the interacting laser beams (Figure 3c, d)

$$\begin{aligned} \omega_p &= \omega_s + \omega_i \\ \vec{k_p} &= \vec{k_s} + \vec{k_i} \end{aligned} \quad (1)$$

where $\vec{k_j}$, $j = p,s,i$, are the wave vectors of the pump, signal, and idler beams.

The host crystal of the parametric process is transparent to the interacting waves, and the amplification takes place without thermal loading of the nonlinear crystal. Exact phase matching condition can be fulfilled only by monochromatic waves. Three beam parameters and three geometrical parameters are involved in a noncollinear OPA (NOPA) process: signal, pump and idler wavelengths, the angle between pump wave-vector and the crystal optical axis (θ), the angle between signal and pump wave-vectors (α), and the angle between signal and idler wave-vectors (β). For a monochromatic noncollinear parametric interaction, three parameters are free-chosen, usually signal wavelength, λ_s, pump wavelength, λ_p, and α angle between signal and pump beams. The idler wavelength (λ_i), θ, and β angles of a NOPA process in a certain nonlinear crystal can be calculated using the phase-matching Eqs. [21]

$$\frac{1}{\lambda_p} = \frac{1}{\lambda_s} + \frac{1}{\lambda_i}$$

$$\frac{n_p(\lambda_p, \theta)}{\lambda_p} \sin\alpha - \frac{n_i(\lambda_i)}{\lambda_i} \sin\beta = 0 \tag{2}$$

$$\frac{n_p(\lambda_p, \theta)}{\lambda_p} \cos\alpha - \frac{n_s(\lambda_s)}{\lambda_s} - \frac{n_i(\lambda_i)}{\lambda_i} \cos\beta = 0$$

Under approximations of small initial signal beam intensity, without input idler beam, and neglected pump beam depletion, the parametric gain is given by [21]

$$G_s(L) = \frac{I_s(L) - I_s(0)}{I_s(0)} = \Gamma^2 \frac{\sinh^2(gL)}{g^2} \tag{3}$$

where L is the length of the nonlinear crystal, $I_s(0)$ is the input signal beam intensity, $I_s(L)$ is the output signal intensity, $g^2 = \Gamma^2 - \left(\frac{\Delta k}{2}\right)^2$, $\Gamma^2 = \frac{2\omega_s\omega_i d_{eff}^2 I_p}{n_s n_i n_p \varepsilon_0 c^3}$, I_p is the pump beam intensity, d_{eff} is the effective nonlinear coefficient, $n_{p,s,i}$ are refractive indexes, ε_0 is the permittivity of free space, c is the speed of light, and $\Delta k = k_p - k_s - k_i$ is the wave-vector mismatch. The full width at half maximum (FWHM) phase-matching bandwidth is usually defined as the spectral range where the parametric gain $G_s(\Delta k)$ is at least 50% from the peak gain obtained in the case of exact phase-matching, $G_s(\Delta k = 0)$ [21]

$$G_s(\Delta k) = \frac{1}{2} G_s(\Delta k = 0) \quad (4)$$

Broad gain bandwidth can be obtained if, near the exact phase-matching condition, the wave-vector mismatch slowly varies depending on the signal wavelength. The Δk phase mismatch can be represented by Taylor series around the phase-matching signal frequency ω_{s0} [22]

$$\begin{aligned}\Delta k = \Delta k^{(0)} + \left(\frac{\partial \Delta k}{\partial \omega_s}\right)_{\omega_{s0}} d\omega_s + \frac{1}{2!}\left(\frac{\partial^2 \Delta k}{\partial \omega_s^2}\right)_{\omega_{s0}} (d\omega_s)^2 + \frac{1}{3!}\left(\frac{\partial^3 \Delta k}{\partial \omega_s^3}\right)_{\omega_{s0}} (d\omega_s)^3 + \\ + \frac{1}{4!}\left(\frac{\partial^4 \Delta k}{\partial \omega_s^4}\right)_{\omega_{s0}} (d\omega_s)^4 + \ldots \approx \Delta k^{(0)} - \left(\frac{\partial k_s}{\partial \omega_s} - \frac{\partial k_i}{\partial \omega_i}\right)\Delta\omega - \frac{1}{2!}\left(\frac{\partial^2 k_s}{\partial \omega_s^2} + \frac{\partial^2 k_i}{\partial \omega_i^2}\right)(\Delta\omega)^2 - \\ - \frac{1}{3!}\left(\frac{\partial^3 k_s}{\partial \omega_s^3} - \frac{\partial^3 k_i}{\partial \omega_i^3}\right)(\Delta\omega)^3 - \frac{1}{4!}\left(\frac{\partial^4 k_s}{\partial \omega_s^4} + \frac{\partial^4 k_i}{\partial \omega_i^4}\right)(\Delta\omega)^4 \ldots = \Delta k^{(0)} + \Delta k^{(1)} + \Delta k^{(2)} + \Delta k^{(3)} + \Delta k^{(4)} + \ldots\end{aligned} \quad (5)$$

where $\Delta k = 0$ represents the condition for quasi-monochromatic phase matching; $\Delta k^{(0)} = \Delta k^{(1)} = 0$ is the condition for optical parametric broad gain bandwidth.

An ultra-broad bandwidth (UBB) of phase-matching can be obtained for $\Delta k^{(0)} = \Delta k^{(1)} = \Delta k^{(2)} = 0$. In this case, two more equations must be added to the three-equation system (2) [22]

$$\begin{aligned} v_{gs} &= v_{gi} \cos\beta \\ \frac{\partial^2 k_s}{\partial \omega_s^2} \cos\beta &+ \frac{\partial^2 k_i}{\partial \omega_i^2} - \frac{\sin^2\beta}{v_{gs}^2 k_i} = 0 \end{aligned} \quad (6)$$

where v_{gs} and v_{gi} are group velocities of signal wave and idler wave, respectively.Particularly in the case of high-energy laser pulse amplification, only a couple of existing high-energy lasers are suitable for OPA pumping. For this reason, usually the pump laser wavelength λ_p represents the free-chosen parameter of the OPA process. For a certain nonlinear crystal, the other five parameters, including signal central wavelength, are deduced from the five-equation system comprising Eqs. (2) and (6).

UBBs of more than 100 nm, able to support amplification of sub-10-fs laser pulses, can be obtained in nonlinear crystals [22], like potassium dideuterium phosphate (DKDP) and BBO. Ultra-broad gain bandwidths for BBO and DKDP crystals, pumped by green nanosecond lasers, in NOPA configuration are shown in Figure 4. Gain bandwidths were calculated assuming plane interacting waves, uniform pump intensity distribution, no input idler beam, and negligible pump beam intensity depletion. For both NOPA processes, I considered a flat pump intensity I_p of 1 GW/cm², which can be accepted without damage risk of currently used nonlinear crystals in case of about one-nanosecond pump pulse duration (e.g., the data sheets of the manufacturing company Altechna) [23]. Different lengths were considered for DKDP and BBO crystals, corresponding to similar gain values in the parametric amplification process.

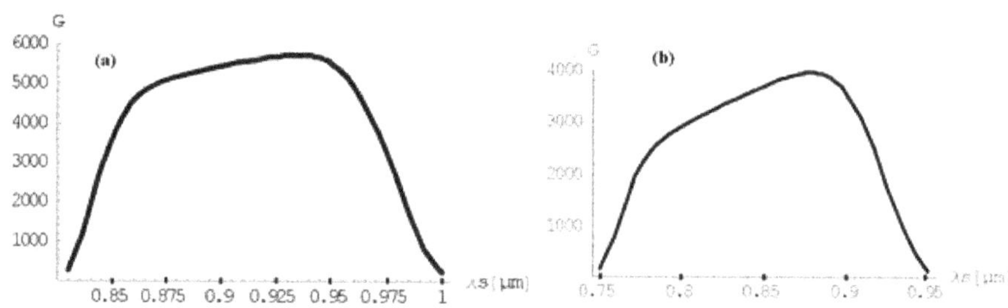

Figure 4. NOPA gain spectra. I_p = 1 GW/cm². (a) 80-mm-long DKDP crystal; $\lambda_{P(DKDP)}$ = 0.527 μm, θ_{DKDP} = 37.0°, α_{DKDP} = 0.92°, λ_{S0} = 900 nm; UBB_{DKDP} ≈ 135 nm. (b) 10-mm-long BBO crystal, $\lambda_{P(BBO)}$ = 0.532 μm, θ_{BBO} = 23.8°, α_{BBO} = 2.4°, λ_{S0} = 0.825 μm; UBB_{BBO} ≈ 150 nm.

The UBB phase-matching of DKDP crystals is centered around λ_{S0} = 900 nm central wavelength, whereas the UBB of BBO crystals is centered in the range of 800 nm wavelength, practically overlapped to the gain bandwidth of Ti:sapphire laser media.

4. OPTICAL PARAMETRIC CHIRPED PULSE AMPLIFICATION

OPCPA was proposed as an alternative solution for the amplification of large bandwidth stretched laser pulses [24] (Figure 5). Drawbacks of the Ti:sapphire CPA, particularly those related to the amplified spectral band narrowing, intensity contrast decrease, and thermal loading, can be overcome in OPCPA laser systems. Signal pulses generated by a broad bandwidth femtosecond oscillator are temporally stretched and synchronized to the pump pulses. Signal and pump pulses have similar durations, usually in the range of picoseconds or nanoseconds. The pump laser wavelength is chosen among the available high-energy green nanosecond lasers, such as frequency-doubled Nd:YAG (532 nm), Nd:glass (527 nm), Yb:YAG (515 nm) lasers. After OPCPA in one or more amplifier stages with nonlinear crystals, enhanced signal pulses can be temporally recompressed to get higher power femtosecond laser pulses.

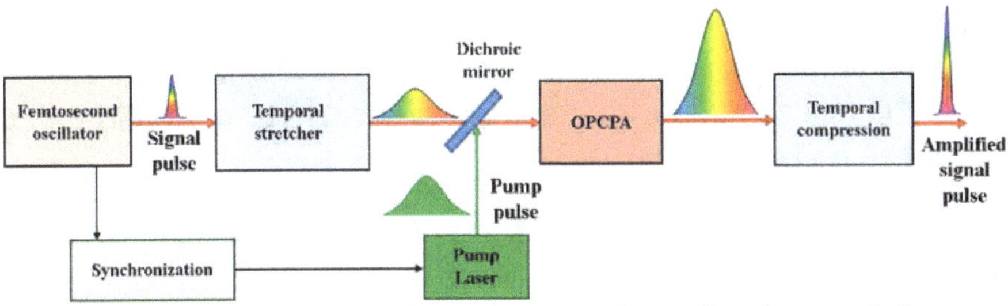

Figure 5. Principle of optical parametric chirped pulse amplification.

Unlike CPA, OPCPA is free from gain narrowing and redshifting effects. Because the host crystal is transparent to the interacting beams, thermal loading is practically absent in the parametric amplification process.

On the other hand, in the case of OPCPA, the spectrum of the amplified laser pulse is sensitive to the angle between signal and pump laser beams. The parametric amplification of each signal spectral component depends on the local instantaneous pump radiation intensity. In order to keep a stable amplified signal spectrum from pulse to pulse, high temporal and spatial stability of the pump beams, as well as very stable experimental setup, are required.

Unlike CPA amplifiers, due to angular constraints between pump and signal wave vectors, imposed by the unique phase-matching geometry, in OPCPA experimental setups usually a single pump laser beam can be used (Figure 6). To amplify broadband chirped laser pulses, laser systems based on noncollinear OPCPA (NOPCPA) configuration, imposed by the conditions of UBB parametric amplification in nonlinear crystals, were developed [3, 25, 26, 27, 28, 29].

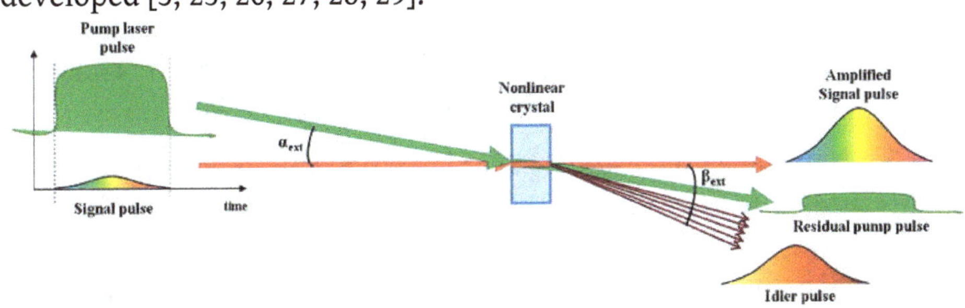

Figure 6. Schematic description of broadband noncollinear OPCPA in nonlinear crystals.

For high-energy final amplifiers of multi-PW laser systems, as much as 10^2–10^3 J pump energy, within ~1 ns pulse duration, is required. It is a real challenge to build a single-beam laser able to deliver the pump pulses for these high-energy OPCPA stages.

5. HIGH-POWER LASER SYSTEMS BASED ON HYBRID AMPLIFICATION

A schematic configuration of a high-power hybrid femtosecond laser system is shown in Figure 7. In the low-energy amplification section, usually called the Front-End (FE) of the laser system, femtosecond pulses generated by a laser oscillator are temporally stretched up to 10 ps – 100 ps. Stretched laser pulses are mainly amplified by OPCPA from the nJ energy level up to 10 mJ – 100 mJ energy range, corresponding to seven to eight orders of magnitude amplification. By high-energy CPA in broad gain bandwidth laser media, chirped laser pulses, stretched in the range of one nanosecond pulse duration, are amplified by three to four orders of magnitude up to 10 J – 100 J and then temporally compressed back to the femtosecond range.

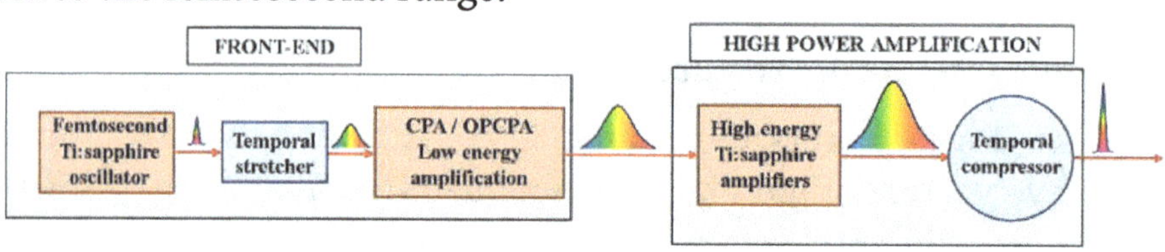

Figure 7. Basic configuration of a hybrid chirped pulse amplification laser system.

OPCPA is considered an appropriate technique in the low energy amplification FE, where large enough nonlinear crystals and good-quality beam ps-ns pump lasers are available. Output FE laser pulses with large spectral bandwidths, recompressible with high intensity contrast, are further amplified in high-energy Ti:sapphire amplifier stages.

In high-power laser systems (HPLS) based exclusively on parametric amplification, the technical bottlenecks move toward very high-energy pump lasers. Pump energy in the kJ range, few nanosecond pulse duration, flat intensity profile, and stable temporal and spatial beam profiles are required for pumping final OPCPA stages.

In femtosecond laser systems, the maximum amplified pulse energy is restricted by the size of available amplifying media. The clear aperture of largest available Ti:sapphire crystals for CPA is smaller than the aperture of some available OPCPA nonlinear crystals, like DKDP, for example. Nevertheless, there is an advantage of Ti:sapphire CPA: for optical pumping of large-aperture Ti:sapphire crystals, several green pump lasers can be used, with output pulse energy of 50–100 J, less restrictive requirements concerning pulse duration, and much higher repetition rate compared to a single-beam kJ pump laser necessary for pumping a high energy OPCPA stage.

Because most of the amplification in hybrid lasers is realized by OPCPA, gain narrowing and ASE effects are attenuated compared to all Ti:sapphire amplifiers. It becomes easier to get high-intensity contrast, high-energy laser pulses, recompressible down to femtosecond pulse duration.

In a hybrid femtosecond pulse amplification system, based on both OPCPA and CPA, a key feature is the matching of the ultra-broad gain bandwidth of the nonlinear crystal to the amplification spectral band of Ti:sapphire laser crystals. In this case, stretched pulses amplified in the laser FE can be directly sent to the Ti:sapphire high-energy amplifiers.

The ultra-broad gain band of DKDP crystals is centered near 900 nm. OPCPA based on DKDP crystals can be used in hybrid femtosecond laser amplifiers. In this case, seed laser pulses must have a broad bandwidth adapted to the ultra-broad phase-matching spectral band of DKDP crystals. In DKDP-OPCPA laser systems equipped with Ti:sapphire broadband femtosecond oscillators, complicated experimental setups were realized to generate broadband laser pulses with the central wavelength shifted near 900 nm [25, 26]. BBO crystals, pumped by frequency-doubled Nd lasers, have a "lucky" ultra-broad phase-matching bandwidth in the range of 800 nm, practically overlapped to the gain bandwidth of Ti:sapphire laser crystals. Due to more than 100 nm phase-matching bandwidth, BBO crystals pumped by green lasers can support the amplification of stretched laser pulses recompressible at sub-10 fs pulse duration [22]. The available few centimeters clear aperture BBO crystals are large enough for OPCPA up to 100 mJ signal pulse energy. For this reason, BBO crystals are frequently used in the FEs of the PW-class hybrid amplification femtosecond laser systems.

Considering the currently available technical solutions, hybrid amplification represents a good choice for the development of petawatt-class femtosecond laser systems.

5.1. PW-class hybrid femtosecond laser systems

A couple of PW-class hybrid femtosecond laser systems are currently worldwide operated, while other 10-PW laser facilities are under development.

A high spatiotemporal quality PW-class laser system has been developed at Advanced Photon Research Center, Japan Atomic Energy Agency [3]. This laser system is based on a double CPA configuration (Figure 8). In the first CPA section, femtosecond laser pulses generated by a Ti:sapphire oscillator are stretched, pre-amplified in Ti:sapphire amplifiers, and temporally recompressed to get mJ-energy output pulses with sub-30 fs duration. To improve the intensity contrast, part of the ASE pedestal of these pulses is removed by a saturable absorber.

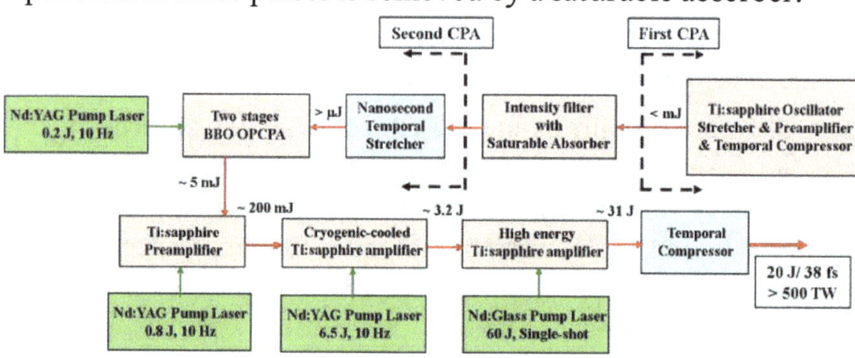

Figure 8. Schematic drawing of a hybrid PW-class laser amplifier based on low-energy OPCPA in BBO crystals and Ti:sapphire high-energy amplification [3].

In the second CPA section, the intensity-filtered pulses, stretched up to ~1 ns pulse duration, are amplified by OPCPA. The conventional regenerative amplifier used in all Ti:sapphire lasers is replaced by a two-stage OPCPA with BBO crystals, pumped by a frequency-doubled Nd:YAG nanosecond laser. Pump and seed pulses are electronically synchronized with a timing jitter of ±0.5 ns. To avoid the parametric fluorescence, the OPCPA stages are operated with relatively high-energy seed pulses in a low gain mode. Seed pulses of ~2.5 μJ are amplified up to ~5 mJ, keeping each BBO stage to less than 100× amplification factor. Hereby, the intensity contrast of amplified pulses in the nanosecond time range is significantly improved, practically with the parametric amplification factor.

Laser pulses are amplified up to 3 J energy level in two Ti:sapphire stages pumped by 10 Hz repetition rate green (532 nm wavelength) Nd:YAG lasers. Final Ti:sapphire amplifier is pumped by a single-shot green (527 nm) nanosecond Nd:glass laser with ~60 J pulse energy. Pump laser beams with smooth homogenized spatial intensity profile are delivered to a large-aperture 80 mm diameter Ti:sapphire laser crystal. A near-homogenous flat-top intensity profile of the amplified pulse beam was obtained. After temporal compression, 20 J/38 fs pulses with more than 0.5 PW peak power and more than 10^{10} intensity contrast in sub-nanosecond temporal range were generated.

1.1 PW laser based on hybrid optical parametric chirped pulse amplification and mixed Nd:glass amplifiers (Figure 9) has been demonstrated at Texas Center of High Intensity Laser Science, Austin, USA [27].

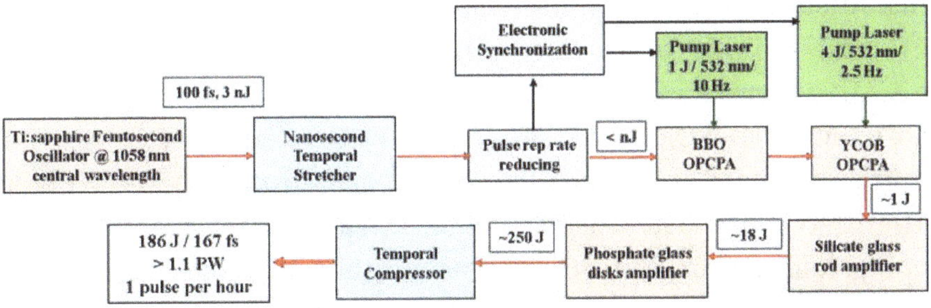

Figure 9. Schematic drawing of a PW laser system based on OPCPA in BBO and YCOB crystals and high-energy amplification in mixed glasses at 1 μm spectral range [27].

Nano-Joule-energy seed pulses with 16 nm FWHM spectrum centered at 1058 nm are generated by a tunable Ti:sapphire laser oscillator. Hundred-femtosecond oscillator pulses are stretched to more than 1 ns pulse duration. Stretched pulses are amplified by approximate nine orders of magnitude in three OPCPA stages to attain ~1 J pulse energy: two stages with pairs of BBO crystals and the last one with a pair of yttrium calcium oxoborate (YCOB) crystals. OPCPA crystals are pumped by frequency-doubled Nd:YAG lasers at 532 nm. The parametric amplification takes place in a near-degeneracy type of interaction, where pump wavelength is about two times shorter than the signal wavelength, which assures a broad enough gain bandwidth to amplify ~30 nm broadband signal pulses. Nanosecond seed and pump pulses are electronically synchronized. Up to 250 J energy, stretched laser pulses are amplified in two flash-lamp pumped Nd:glass amplifiers, with shifted peak gain wavelengths of the Nd-doped glasses. The gain spectral bandwidth of mixed Nd:glass amplifiers is about 15 nm near 1 μm central wavelength. First amplifier stage consists in a 64-mm-diameter silicate rod amplifier. In the second amplifier stage, the laser pulse passes four times through two pairs of 315 mm aperture phosphate disk amplifiers. Finally 186 J, 168 fs compressed pulses, with estimated nanosecond range contrast better than 10^{12} were obtained.

A high-contrast 1.16 PW laser system was developed at Beijing National Laboratory for Condensed Matter Physics, Institute of Physics, by combining low-energy femtosecond optical parametric amplification with high-energy Ti:sapphire amplification [4]. Sub-10 fs pulses, with 4 nJ pulse energy, generated by a Ti:sapphire oscillator are split into two beams. About 70% energy pulse was used as the seed for a Ti:sapphire CPA to obtain 5 mJ, 50 fs pulses, which are frequency-doubled by BBO crystals to generate second harmonic pulses for pumping the two-stage broad bandwidth NOPA. The other 30% of the femtosecond oscillator pulse energy is used as the broadband seed, at 800 nm central wavelength, for the first NOPA. The optical synchronization of the signal and the pumping femtosecond pulses was accurately controlled by a Herriot telescope delay line. This way, large bandwidth femtosecond pulses are amplified to ~26 μJ energy without any temporal stretching in the two-stage NOPA with BBO crystals. The parametric amplification process which involves the signal and pump pulses occurs on few 10 fs timescale. The background noise beyond this time range cannot be amplified, and the amplified signal pulse contrast is improved by a factor equal to the parametric gain. In the second CPA stage, the clean signal pulses were stretched to about 600 ps. The medium energy Ti:sapphire amplifiers are pumped by 532 nm wavelength nanosecond Nd:YAG lasers running at 1

Hz repetition rate. The last high-energy amplifier consists in an 80-mm-diameter, 40-mm-thickness Ti:sapphire disk, pumped by a nanosecond Nd:glass laser, 120 J at 527 nm wavelength, 1 pulse/20 minutes repetition rate. After temporal compression, more than 32 J pulse energy at ~28 fs pulse duration, corresponding to a peak power up to 1.16 PW, with enhanced intensity contrast ratio of 10^{10}, has been obtained.

In the frame of the French project *Apollon*, an optically synchronized OPCPA with picosecond pulses was proposed for the Front-End configuration of a 10 PW laser system [28]. A similar solution was considered for the 2 × 10 PW laser system of the ELI-NP laser facility from Bucharest-Magurele [29].

5.2. High-power laser system of the ELI-NP research facility

High-power laser system (HPLS) of the ELI-NP laser facility, developed by Thales Optronique, consists in a two-arm 10 PW peak power femtosecond laser amplifier. ELI-NP HPLS combines the advantages of a high amplification factor FE, based on OPCPA at low energy level, with the high energy amplification in large-size Ti:sapphire crystals pumped by high-energy frequency-doubled Nd:YAG and Nd:glass lasers [29]. For each arm, two additional output beams of 100 TW pulse peak power at 10 Hz repetition rate and 1 PW at 1 Hz are available. The schematic drawing of the ELI-NP 2 × 10 PW laser system is shown in Figure 10.

Pump pulse is obtained by amplifying a pJ-energy pulse, generated at the edge of the fs oscillator spectral bandwidth, with the central wavelength of 1064 nm and ~10 nm spectral bandwidth. In the first diode-pumped Ytterbium-doped fiber amplifier, pulse energy is increased to the nJ range. After spectral filtering in a Fiber Bragg Grating, the spectral bandwidth of the pulse is reduced to less than 0.1 nm bandwidth and its energy decreases in the range of 10 pJ. After amplification in the second Ytterbium-doped fiber amplifier, near-Fourier-transform-limited pulses of ~1 nJ energy and ~25 ps pulse duration are amplified in bulk Nd:YAG amplifiers and frequency-doubled in a Lithium Triborate (LBO) crystal to get more than 80 mJ pump pulse energy at 532 nm wavelength and 10 Hz repetition rate.

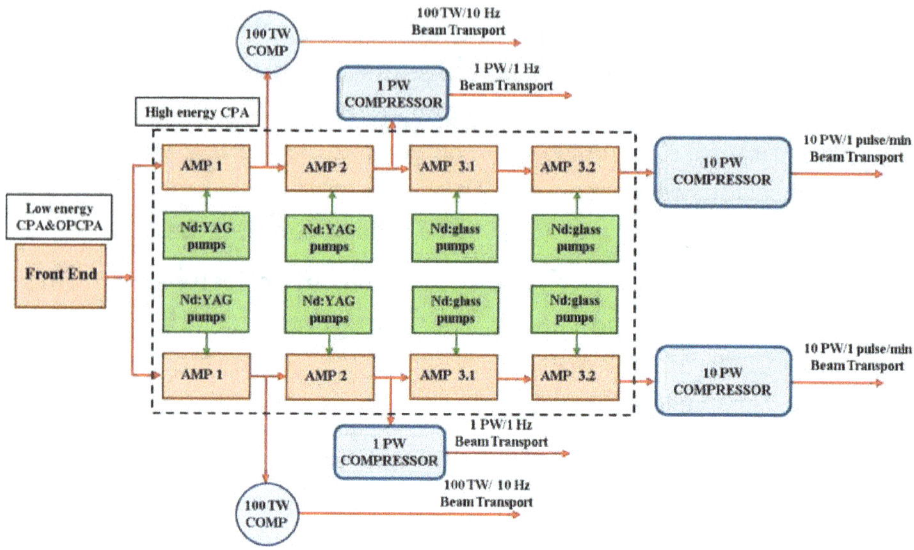

Figure 10. Schematic drawing of the 2 × 10 PW ELI-NP femtosecond laser system.

FE is based on OPCPA with optically synchronized seed and pump pulses of 20 to 25 ps duration (Figure 11). Seed and pump pulses for OPCPA are created by amplifying two output pulses generated by an ultra-broad bandwidth Ti:sapphire femtosecond oscillator (Venteon Company).

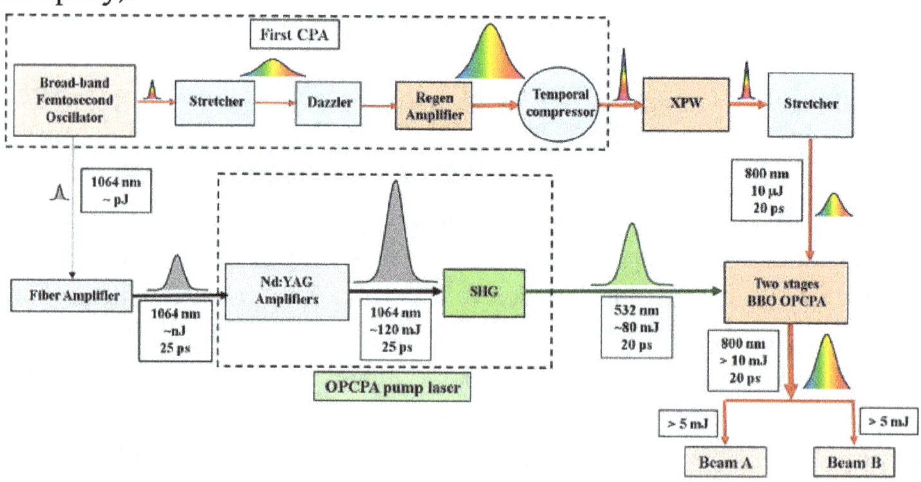

Figure 11. ELI-NP HPLS front-end based on optically synchronized OPCPA with BBO crystals.

The broadband seed pulses of nJ pulse energy at ~800 nm central wavelength are temporally stretched to 100 ps in a diffraction grating stretcher. An acousto-optic programmable dispersion filter (AOPDF), Dazzler type, is used to compensate for high-order phase distortions [9]. Stretched pulses are amplified to the mJ level in a Ti:sapphire regenerative amplifier and recompressed in the range of few 10 fs pulse duration. Femtosecond pulses amplified in the first CPA are intensity filtered and spectrally broadened by XPW generation in two barium fluoride (BaF_2) crystals. The femtosecond pulses with improved intensity contrast are stretched to ~20 ps pulse duration in a bulk glass stretcher. Broad bandwidth picosecond pulses are amplified in a two-stage NOPCPA with BBO crystals. To preserve a high intensity contrast during the parametric amplification process, the parametric fluorescence is hampered by keeping a relatively low parametric gain, less than 100× amplification factor for each NOPCPA stage. Outside the temporal window of the parametric process, the intensity contrast is improved by a numerical factor equal to the parametric gain.

By combining XPW and NOPCPA, the intensity contrast of recompressed pulses after FE amplification was improved by at least six orders of magnitude. More than 10^{12} ASE intensity contrast has been estimated in the few 10 ps range before the main femtosecond pulse. Spectral bandwidth narrowing and redshifting effects, specific to all Ti:sapphire amplification, were drastically reduced in the FE. Picosecond pulses with more than 70 nm FWHM spectral bandwidth were obtained at the FE output [30, 31]. Amplified pulses of ~10 mJ energy, split into two equal-energy beams, represent the seed pulses for A and B HPLS Ti:sapphire amplification arms.

After stretching to ~1 ns duration, laser pulses will be amplified up to the level of few 100 J in the all Ti:sapphire second CPA system (Figure 10). Ti:sapphire amplifiers AMP 1 are pumped by frequency-doubled Nd:YAG lasers at 10 Hz repetition rate. More than 4 J energy of the amplified chirped pulses can be obtained. By temporal compression of these pulses, 100 TW beams are generated in each arm of the HPLS. The next amplifier, AMP 2, is pumped by Nd:YAG lasers at 1 Hz repetition rate to get ~36 J pulse energy required for the generation of 1-PW temporally compressed laser pulses. In the last amplification stages, AMP 3.1 and

AMP 3.2, Ti:sapphire crystals are pumped by 100-J energy frequency-doubled Nd:glass lasers (Atlas 100, Thales Optronique Company) at 1 pulse/min repetition rate [32].

To attain the 10-PW peak power with as low as possible pulse energy, a large spectral bandwidth of laser pulses must be preserved throughout all amplification process. To compensate for redshifting and gain narrowing effects in the high-energy Ti:sapphire amplifiers, the output spectrum will be managed using reflective filters for spectrum shaping at the input of AMP 1, AMP 2, and AMP 3 amplifiers, similar to the solution proposed for the 10-PW *Apollon* laser in a previously published work [7]. Final spectral bandwidth as broad as 60 nm with the central wavelength of ~815 nm is expected, compared to ~35 nm bandwidth at ~845 nm central wavelength, calculated without spectrum control. The improved spectral bandwidth theoretically allows the generation of recompressed pulses as short as 15 fs. To secure the 10-PW peak power of the HPLS in case of 22- to 23 fs duration of the temporally recompressed pulses, more than 300 J energy of amplified chirped pulses could be obtained by full energy pumping of the last amplifier stage, AMP 3.2. Considering a safe laser fluence of 1–1.5 J/cm^2, Ti:sapphire crystals with clear aperture diameter in the range of 160–200 mm are necessary for the last amplifier stages.

Main specifications of the ELI-NP HPLS are summarized in the Table 1.

Table 1. Main specifications of ELI-NP HPLS.

Laser beam parameter	Estimated value
Laser pulse peak power	≥ 10 PW
Estimated pulse duration	15–25 fs
Estimated pulse energy	150–250 J
Repetition rate	1 pulse/min
Intensity contrast	≥ 10^{12}

To reach as high as 10^{23} W/cm^2 focused beam intensity, the 10-PW laser beam must be tightly focused in a few micrometers spot. Wavefront distortions, produced mainly by the thermal loading of Ti:sapphire amplifiers, give rise to focal intensity profile aberrations. Their accumulation occurs through the increase in the size of the focal point and the decrease in the energy content in the primary spot. The Strehl ratio, which quantifies the peak intensity compared to an ideal flat wavefront, can decline to values below 0.2 [10, 33]. Adaptive optics is necessary for precise and consistent laser beam focussing by controlling and correcting wavefronts. Following the high-energy amplifiers, deformable mirrors will be put in each arm of the HPLS amplification system. Anticipate output beam wavefronts that possess a Strehl ratio exceeding 0.8 and focused spots that are nearly diffraction-limited.

6. CONCISE SUMMARY

Hybrid high-power femtosecond laser systems merge the benefits of ultra-wide bandwidth optical parametric chirped pulse amplification (OPCPA) in nonlinear crystals with the chirped pulse amplification (CPA) approach in large-sized titanium-doped sapphire (Ti:sapphire) crystals. Nonlinear crystals such as BBO and DKDP, when pumped by green lasers, can achieve ultra-broad gain bandwidths of around 150 nm using noncollinear optical parametric chirped pulse amplification. One important characteristic of hybrid amplification lasers is their ability to adjust the phase-matching bandwidth of OPCPA nonlinear crystals to match the gain bandwidth of laser-amplifying media, such as Ti:sapphire crystals and Nd:doped glasses. The spectral bandwidths of BBO crystals for phase-matching and Ti:sapphire laser crystals for gain are overlapping. OPCPA is used to achieve significant amplification in hybrid femtosecond laser systems, often resulting in amplification orders of magnitude. The narrowing effect and intensity pedestal of the gain are much reduced compared to all other Ti:sapphire amplifiers. Obtaining high intensity contrast, broad spectrum bandwidth, and high-energy femtosecond laser pulses becomes more convenient. Hybrid femtosecond laser amplifiers, utilizing optical parametric chirped pulse amplification (OPCPA) in beta-barium borate (BBO) crystals and chirped pulse amplification (CPA) in titanium-doped sapphire (Ti:sapphire) crystals, can generate high-power laser pulses with a pulse width of 10 femtoseconds.

The ELI-NP research facility is currently constructing a hybrid amplification femtosecond laser system with a power of 2 × 10 PW. The Front-End utilizes optically synchronized picosecond pulses generated by an optical parametric chirped pulse amplification (OPCPA) system in beta barium borate (BBO) crystals. The ELI-NP laser Front-End produced picosecond stretched pulses with an energy of around 10 millijoules. These pulses had a spectral bandwidth wider than 70 nanometers. The ELI-NP high-power laser system is capable of producing 10-PW pulses with a duration of approximately 20 fs. These pulses are achieved using high-energy chirped pulse amplification in large aperture Ti:sapphire crystals, followed by temporal recompression. At the output of the system, the pulses are expected to have an intensity contrast of more than 10^{12} picoseconds ASE.

AT A GLANCE

Hybrid femtosecond lasers integrate the process of chirped pulse amplification (CPA) in laser medium with optical parametric chirped pulse amplification (OPCPA) in nonlinear crystals. Nonlinear crystals can be used to achieve gain bandwidths as wide as 150 nm by noncollinear optical parametric chirped pulse amplification. Thus, laser pulses that can be compressed to a pulse duration of less than 10 femtoseconds can be amplified in crystals such as beta-barium borate (BBO) and potassium dideuterium phosphate (DKDP). The phase-matching bandwidth of beta-barium borate crystals, which is about 800 nm wavelength, significantly overlaps with the gain bandwidth of Ti:sapphire laser crystals when pumped by green nanosecond lasers. Optical parametric chirped pulse amplification in beta-barium borate crystals at a low-energy level in the laser system Front-End (FE), when paired with high-energy chirped pulse amplification in Ti:sapphire crystals, offers a sophisticated solution for femtosecond laser systems of petawatt-class. Several globally advanced hybrid amplification high-power femtosecond laser systems are showcased. This text describes the configuration and properties of the output beam of the hybrid amplification petawatt laser at the Extreme Light Infrastructure: Nuclear Physics (ELI-NP) facility.

REFERENCES

1. Strickland D, Mourou G. Compression of amplified chirped optical pulses. Optics Communication. 1985;56:219-221
2. Sung JH, Lee SK, TJ Y, Jeong TM, Lee J. 0.1 Hz 1.0 PW Ti: sapphire laser. Optics Letters. 2010;35:3021-3023

3. Kiriyama H, Michiaki M, Nakai Y, Shimomura T, Sasao H, Tanaka M, Ochi Y, Tanoue M, Okada H, Kondo S, Kanazawa S, Sagisaka A, Daito I, Wakai D, Sasao F, Suzuki M, Kotakai H, Kondo K, Sugiyama A, Bulanov S, Bolton PR, Daido H, Kawanishi S, Collier JL, Hernandez-Gomez C, Hooker CJ, Ertel K, Kimura T, Tajima T. High-spatiotemporal-quality petawatt-class laser system. Applied Optics. 2010;49:2105-2115

4. Wang Z, Liu C, Shen Z, Zhang Q, Teng H, Wei Z. High-contrast 1.16 PW Ti: sapphire laser system combined with a doubled chirped-pulse amplification scheme and a femtosecond optical-parametric amplifier. Optics Letters. 2011;36:3194-3196

5. TJ Y, Lee SK, Sung JH, Yoon JW, Jeong TM, Lee J. Generation of high-contrast, 30 fs, 1.5 PW laser pulses from chirped-pulse amplification Ti: sapphire laser. Optics Express. 2012;20:10807-10815

6. Chu Y, Liang X, Yu L, Xu Y, Xu L, Ma L, Lu X, Liu Y, Leng Y, Li R, Xu Z. High-contrast 2.0 Petawatt Ti: sapphire laser system. Optics Express. 2013;21:29231-29239

7. Giambruno F, Radier C, Rey G, Chériaux G. Design of a 10-PW (150J/15 fs) peak power laser system with Ti: sapphire medium through spectral control. Applied Optics. 2011;50:2617-2621

8. Jullien A, Canova L, Albert O, Boschetto D, Antonucci L, Cha YH, Rousseau JP, Chaudet P, Cheriaux G, Etchepare J, Kourtev S, Minkovski N, Saltiel SM. Spectral broadening and pulse duration reduction during cross-polarized wave generation: influence of the quadratic spectral phase. Applied Physics B: Lasers and Optics. 2007;87:595-601

9. Verluise F, Laude V, Huignard JP, Tournois P, Migus A. Arbitrary dispersion control of ultrashort optical pulses with acoustic waves. Journal of the Optical Society of America B: Optical Physics. 2000;17:138-145

10. Akahane Y, Ma J, Fukuda Y, Aoyoma M, Kiriyama H, Sheldakova JV, Kudryashov AV, Yamakawa K. Characterization of wave-front corrected 100 TW, 10 Hz laser pulses with peak intensities greater than 10^{20} W/cm^2. Review of Scientific Instruments. 2006;77: 023102

11. Yanovsky V, Chvykov V, Kalinchenko G, Rousseau P, Planchon T, Matsuoka T, Maksimchuk A, Nees J, Cheriaux G, Mourou G, Krushelnick K. Ultra-high intensity 300-TW laser at 0.1 Hz repetition rate. Optics Express. 2008;16:2109-2114

12. Hong KH, Hou B, Nees JA, Power E, Mourou GA. Generation and measurement of $>10^8$ intensity contrast ratio in a relativistic kHz chirped-pulse amplified laser. Applied Physics B: Lasers and Optics. 2005;81:447-457

13. Fourmaux S, Payeur S, Buffechoux S, Lassonde P, St-Pierre C, Martin F, Kieffer JC. Pedestal cleaning for high laser pulse contrast ratio with a 100 TW class laser system. Optics Express. 2011;19:8486-8497

14. Minkovski N, Petrov GI, Saltiel SM, Albert O, Etchepare J. Nonlinear polarization rotation and orthogonal polarization generation experienced in a single-beam configuration. Journal of the Optical Society of America B: Optical Physics. 2004;21:1659-1664

15. Jullien A, Albert O, Burgy F, Hamoniaux G, Rousseau JP, Chambaret JP, Augé-Rochereau F, Chériaux G, Etchepare J, Minkovski N, Saltiel SM. 10^-10 temporal contrast for femtosecond ultraintense lasers by cross-polarized wave generation. Optics Letters. 2005;30:920-922

16. Jullien A, Rousseau JP, Mercier B, Antonucci L, Albert O, Chériaux G, Kourtev S, Minkovski N, Saltiel SM. Highly efficient nonlinear filter for femtosecond pulse contrast enhancement and pulse shortening. Optics Letters. 2008;33:2353-2355

17. Dromey B, Kar S, Zepf M, Foster P. The plasma mirror—A subpicosecond optical switch for ultrahigh power lasers. Revue of Scientifical Instruments. 2004;75:645-649

18. Kapteyn HC, Murname MM, Szoke A, Falcone RW. Prepulse energy suppression for high-energy ultrashort pulses using self-induced plasma shuttering. Optics Letters. 1991;16:490-492

19. Musgrave I, Shaikh W, Galimberti M, Boyle A, Hernandez-Gomez C, Lancaster K, Heathcote R. Picosecond optical parametric chirped pulse amplifier as a preamplifier togenerate high-energy seed pulses for contrast enhancement. Applied Optics. 2010;49:6558-6562

20. Koechner W. Solid-State Laser Engineering. New York: Springer Science+Business Media; 2006 ISBN-10: 0-387-29094-X

21. Cerullo G, De Silvestri S. Ultrafast optical parametric amplifiers. Review of Scientific Instruments. 2003;74. DOI: 10.1063/1.1523642

22. Dabu R. Very broad gain bandwidth parametric amplification in nonlinear crystals at critical wavelength degeneracy. Optics Express. 2010;18:11689-11699

23. www.altechna.com

24. Dubietis A, Jonušauskas G, Piskarskas A. Powerful femtosecond pulse generation by chirped and stretched pulse parametric amplification in BBO crystal. Optics Commun. 1992;88:437-440

25. Tang Y, Ross IN, Hernandez-Gomez C, New GHC, Musgrave I, Chekhlov OV, Matousek P, Collier JL. Optical parametric chirped-pulse amplification source suitable for seeding high-energy systems. Optics Letters. 2008;33:2386-2388

26. Lozhkarev VV, Freidman GI, Ginzburg VN, Katin EV, Khazanov EA, Kirsanov AV, Luchinin GA, Mal'shakov AN, Martyanov MA, Palashov OV, Poteomkin AK, Sergeev AM, Shaykin AA, Yakovlev IV. Compact 0.56 Petawatt laser system based on optical parametric chirped pulse amplification in KD*P crystals. Laser Physics Letters. 2007;4:421-427

27. Gaul EW, Martinez M, Blakeney J, Jochmann A, Ringuette M, Hammond D, Borger T, Escamilla R, Douglas S, Henderson W, Dyer G, Erlandson A, Cross R, Caird J, Ebbers C, Ditmire T. Demonstration of a 1.1 petawatt laser based on a hybrid optical parametric chirped pulse amplification/mixed Nd: glass amplifier. Applied Optics. 2010;49:1676-1681

28. Chambaret JP. The Extreme Light Infrastructure Project ELI and its prototype APOLLON/ILE. LEI Conference, 21 October 2009, Brasov, Romania.

29. http://www.eli-np.ro/documents/ELI-NP-WhiteBook.pdf

30. Chalus O, Pellegrina A, Casagrande O, Derycke C, Boudjemaa L, Simon-Boisson C, Laux S, Lureau F, Sanchez D, Biegert J, Ahrens J, Binhammer T, Prochnow O, Rausch S. High Contrast Broadband

Seeder for Multi-PW Laser System. CF-P 19, CLEO-Europe EQEC, 21–25 June 2015, Munchen

31. Lureau F, Laux S, Casagrande O, Chalus O, Duvochelle PA, Herriot S, Matras G, Radier C, Boudjemaa L, Simon-Boisson C, Dabu R, Dancus I, Ursescu D. Design and initial results of 10 PW laser for ELI-NP. CF-P 20, CLEO-Europe EQEC, 21–25 June 2015, Munchen
32. Casagrande O, Derycke C, Soujae A, Ramos P, Boudjemaa L, Simon-Boisson C, Laux S, Lureau F. High energy pump laser for Multi-Petawatt laser. CA-3.2, CLEO-Europe EQEC, 21–25 June 2015, Munchen
33. Fourmaux S, Payeur S, Alexandrov A, Serbanescu C, Martin F, Ozaki T, Kudryashov A, Kieffer JC. Laser beam wavefront correction for ultra-high intensities with the 200 TW laser system at the Advanced Laser Light Source. Optics Express. 2008;16:11987-11993

CHAPTER-4

NEXT-GENERATION LASER SYSTEMS WITH EXCEPTIONALLY HIGH PEAK AND AVERAGE POWER

1. INTRODUCTION: AN OVERVIEW

Currently, the sole method to achieve an exceptionally high level of energy concentration in a compact space is by the use of ultra-high peak power laser pulses, which can be focused to an intensity of 1022 W/cm^2 and higher. The potential to achieve high average power (high repetition rate) in pulse laser systems makes them highly valuable tools for scientific investigation, as well as potentially useful for industrial applications. Pulsed lasers necessitate far less energy to achieve ultra-high powers in comparison to continuous wave (CW) regimes. The situation became quite favorable when the Q-switching [1] and mode-locking [2] methods of laser operation were developed (Figure 1). This enabled the generation of laser pulses of increasingly shorter durations by utilizing mode-locking oscillators. Initially, nanosecond-level pulses (10^{-9} s) were achieved, followed by picosecond (10^{-12} s) and subsequently femtosecond (10^{-15} s) pulse durations. The lasers, utilizing these technologies and employing the master oscillator power amplifier (MOPA) design, have successfully attained power levels in the range of megawatts (MW) and gigawatts (GW), equivalent to 1011–1014 watts per square centimeter (W/cm^2) after focusing. The capacity to achieve these elevated levels of power has opened up possibilities for utilizing these systems in other emerging and highly promising domains, such as charged particle acceleration [3] and the confinement of fusion nuclear reactions by inertial confinement (ICF) [4]. The further development of these laser systems led to the creation of the impressive National Ignition Facility (NIF) laser in the United States [5], and the Megajoule in Europe [6].

The previous system, which emits energy through 192 laser beams with a total output of around 4 megajoules, is capable of delivering a power of up to 500 terawatts to the target. The laser spans an area equivalent to three football fields (refer to Figure 2a) and is comprised of two laser bays, one of which is depicted in Figure 2b.

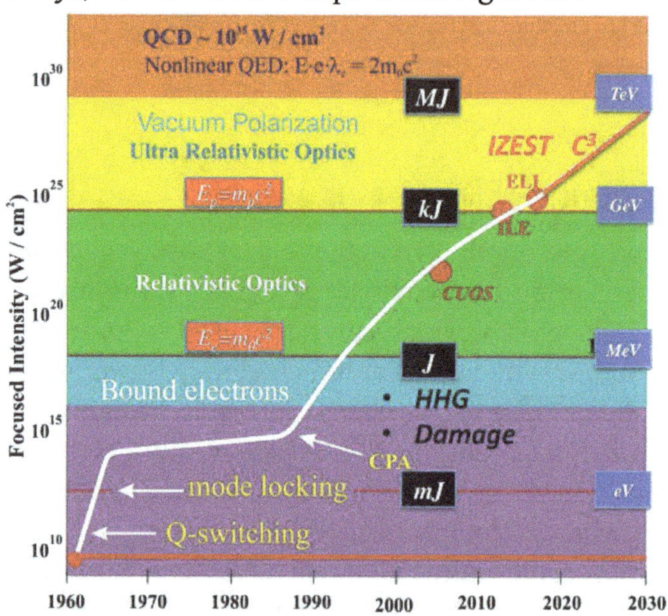

Figure 1. Intensity evolution since the first laser demonstration in 1960, with the different regimes of optics and electrodynamics. Black boxes (joules) indicate typical laser energies. Blue boxes (electron volts) indicate typical laser-plasma accelerated particle energies. (Courtesy G. Mourou.) http://www.spie.org/newsroom/4221-exploring-fundamental-physics-at-the-highest-intensity-laser-frontier?highlight=x2404&ArticleID=x88664&SSO=1 (QCD, quantum chromod-ynamics; QED, quantum electrodynamics; E, electric field; e, electron charge; λ_c, compton wavelength; m_0, electron mass; c, speed of light. E_p, proton energy; m_p, proton mass. E_e, electron energy; C^3, cascaded conversion compression; ELI, extreme light infrastructure; ILE, institut de la lumiere extreme; CUOS, Center for Ultrafast Optical Science University of Michigan; HHG, high harmonic generation; CPA, chirped pulse amplification).

These multichannel facilities require a large area because the damage threshold of the optical elements is typically restricted to 10 J/cm^2 for pulse durations in the nanosecond range. Further augmenting the strength is linked to an intolerable enlargement of the laser systems, known as gigantism.

However, it took more than 20 years after the invention of the Q-switch for the next significant advancement in increasing laser powers. This advancement was called chirped pulse amplification (CPA) technology and was proposed in the mid-1980s. It was associated with a significant reduction in pulse duration, resulting in an increase in peak power. This is illustrated in Figure 1 [7, 8]. The primary objective of the CPA approach is to minimize the dimensions of optical elements and amplifier crystals. Attempting to directly amplify the brief laser pulse can quickly result in unwanted nonlinear aberrations in the gain media and other optical components. As a result, the laser beam undergoes self-focusing, causing an increase in intensity that ultimately leads to the deterioration of the optical components. Thus, it is necessary to enlarge the beam width and the apertures of the amplifiers and other optics in order to prevent these issues. However, in order to prevent the intensity from above the self-focusing threshold, it is necessary to have very large apertures for the optical elements, such as several meters in diameter for petawatt output power.

(a) (b)

Figure 2. (a) NIF laser and target area building; (b) laser bay 2, one of NIF's two laser bays (from https://www.theguardian.com/environment/gallery/2009/may/28/national-ignition-facility-fusion-energy#img-7. http://archive.boston.com/bigpicture/2010/10/the_national_ignition_facility.html).

The progressive concept of the CPA (Chirped Pulse Amplification) was implemented for the optical diapason, providing an alternative method to enhance pulse energy. The suggestion was made to increase the pulse duration instead of extending the size of the transverse beam, in order to achieve greater amplification. This was then followed by the process of pulse compression. Initially, this concept was implemented in the microwave range for radar technology throughout the 1960s. The stretching and compression of the laser pulse were proposed using various methods, including compression in dispersive media [9], multilayer film interferometers [10], and the most extensively studied and productive method today, stretching and compression with a pair of diffraction gratings [11]. In 1985, the initial amplification and compression of a pulse that was chirped in a dispersive media (specifically, 1.4 km of optical single mode fiber) was achieved using a pair of diffraction gratings [7, 8]. Nevertheless, the higher orders dispersion of the optical fiber caused a distortion in the chirped pulse, which posed a challenge in compensating for it using the grating compressor. Consequently, the output pulse duration was restricted to a few picoseconds.

The most successful implementation of this concept (Martínez-type stretcher setup with a telescope positioned between the gratings [12]) is depicted in Figure 3. In this configuration, both the stretcher and compressor were constructed using diffraction gratings as their foundation. The primary evidence can be readily elucidated. A short pulse should have a broad spectrum as a result of the Heisenberg uncertainty principle, meaning it is composed of several harmonics. By utilizing diffraction, we have the ability to redirect various components of this spectrum, known as harmonics, along separate optical routes. This allows us to achieve distinct group delays for each harmonic, ultimately steering them towards the same direction. As a result, the pulse duration expands accordingly. If the stretcher directs harmonics with longer wavelengths along shorter optical channels (positive or normal dispersion), then the compressor must achieve the opposite effect (negative or anomalous dispersion). Due to the utilization of diffraction gratings in both the stretcher and the compressor, the stretching and compression functions will be in close proximity, allowing for compensation of higher-order dispersion up to the 4th order.

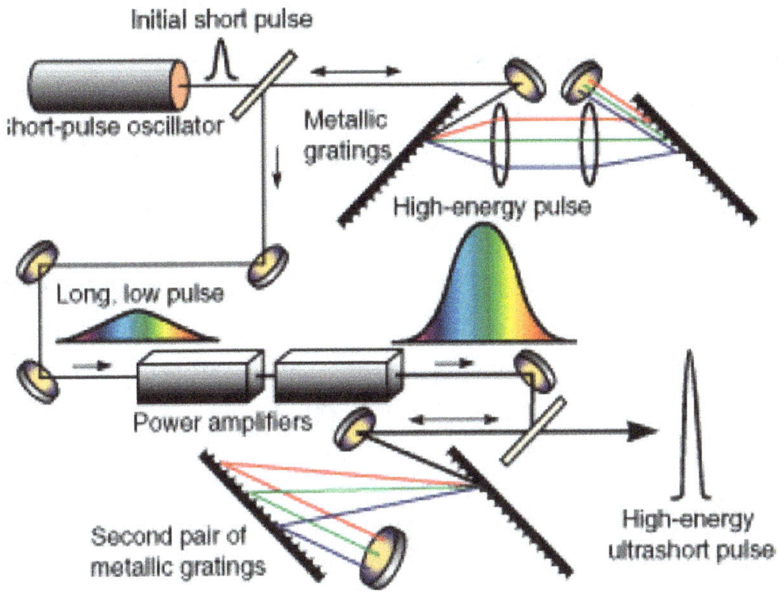

Figure 3. Simplified layout of the CPA laser system (from http://www.llnl.gov/str/Petawatt.html).

Figure 3 displays a simplified arrangement of the classical CPA laser system. The brief pulse emitted by the mode-locked oscillator is typically enlarged by a factor of 4-5 when it passes through the stretcher. Additionally, the low-intensity pulse is greatly magnified in the series of amplifiers by multiple orders of magnitude, and subsequently compressed back to its original brief duration. Various laser active media, including dye, Nd:YAG, Nd:Glass, and others, were employed as oscillators and amplifiers in the CPA systems. The Ti:Sapphire (Ti:Sa) crystal was the most favored among them because to its wide emission spectra bandwidth (FWHM ~ 200 nm), high thermal conductivity, and mechanical toughness. If the Ti:Sa oscillators [13] with a few femtosecond mode-locked pulse and amplifiers are used, this configuration can produce pulse durations shorter than 10 femtoseconds. In addition to laser amplifiers [14], optical parametric amplifiers (OPAs) have been introduced in CPA configurations [15].

Utilizing this technology, scientists have created laser systems that can produce output power at the level of 100 s terawatts and intensity of 1019 W/cm2 (refer to Figure 1). However, like any other technology, chirped pulse amplification (CPA) has certain limitations on the ongoing enhancement of its parameters, the most significant of which is the restriction on the output energy.

1.1. Limitation on the output energy

As mentioned above, CPA laser systems have reached sub-petawatt output powers with the energy as low as several tens of Joules [16]. However, for that new generation of the ultra-high power lasers, the modest energy will no longer be enough, and the kJ-level has to be reached setting the next milestone at tens or even hundreds of petawatt [17, 18]. The modern ability to stretch pulse duration is restricted at the few orders by the existing grating technology, so further energy increase requires once again the enlargement of amplifier apertures due to the damage threshold. There are two candidates for the final amplifier of the ultra-high CPA systems: the optical parametrical amplifiers [15] and the laser amplifiers. Both of them possess their own advantages and shortcomings. OPA has several attractive properties as the gain bandwidths large enough to support light pulses as shorter as 10 fs; low amplifier heating due to "Optical cooling," which leads to the potential of a high-repetition-rate; large high-quality nonlinear amplifier crystals; high pulse contrast due to the absence of amplified spontaneous emission (ASE) outside the pump pulse duration. At the same time, OPA possesses low efficiency (below 20%); there are severe requirements on the pump beam quality, such as limitation on the spatial fluctuations in amplitude; short pump pulses (below 1 ns), in order to match with the stretched signal pulses for higher efficiency and their precise matching in space and time with a ps accuracy; and so on [15].

The laser amplifiers are free of most of these restrictions, but possess problems of their own, which include severe losses due to th disk shape of the crystals and very high transverse gain and so the high transverse ASE (TASE), and possible parasitic generation (TPG). TASE as well as TPG leads to significant depletion of the inverted population, and thus the stored energy. The ASE from the pumped Ti:Sa crystal before and after the threshold of the parasitic generation is demonstrated in Figure 4 [16]. The absence of ASE from the significant part of crystal demonstrates the reduction of the population inversion and so loses the stored energy. Because of this, enlarging the aperture of the amplifier fails to be an unlimited means of obtaining more output energy under the constraints imposed by the damage threshold limit. Therefore, in this case, the main limitation that arises on the path toward ultra-high output power and intensity is the restriction on the pumping and extraction energy imposed by TASE and TPG within the booster amplifier volume. As a result, the suppression of parasitic generation is a very important task that has been solved for the next generation of the ultra-high power laser systems. The technology allowed to solve this bottleneck problem will be discussed below in this chapter.

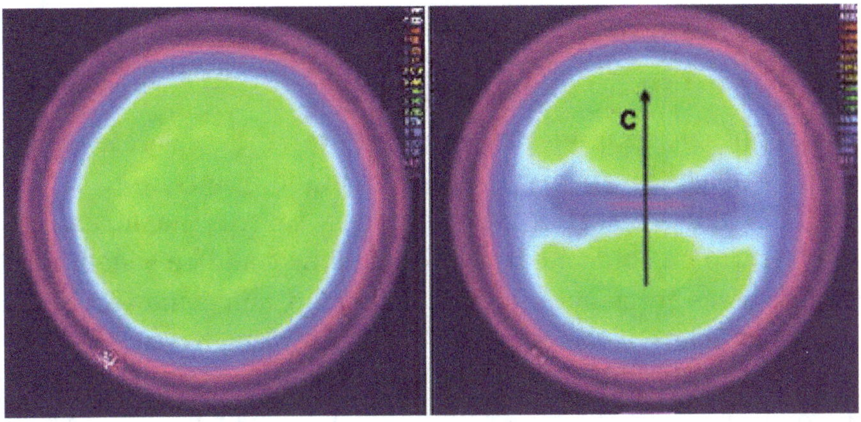

Figure 4. ASE from the pumped Ti:Sapphire crystal before and after threshold of parasitic generation [16].

1.2. Gain narrowing during amplification

The making of the output short pulse shorter and keeping the same energy paved another way to the next milestone of the ultra-high peak power. Nevertheless, pulse duration of CPA lasers is strongly limited by gain narrowing of the pulse spectra in the gain medium of the multipass amplifiers (to around 30 fs of 100 TW–1 PW lasers based on Ti:Sa). Different approaches were entertained to overcome this limitation. Among them, there is a gain narrowing control by introduction of the thin film etalon into the front-end regenerative amplifier cavity. Impressive results were reached with this technique (pulse duration—16 fs with spectral width—72 nm) [19], but this method still suffers some restrictions, such as losses of the energy, a finite possibility of the broadening of the spectra due to the fluorescence spectrum limitation of the amplifier crystal, further gain narrowing in the buster and final amplifiers, as well as the limitation of the spectral transmission of the grating compressor.

The application of concept of optical-parametrical chirped pulse amplification (OPCPA) for broadband pulse allowed to reach a pulse duration of around 10 fs [20], but extremely severe requirements on the parameters of the pump laser which discussed above restrict output energy only to a few mJs. This way, an adequate method for shortening high-energy pulses has not yet been found. Two promising methods of the saving and restoring spectral bandwidth will be discussed also in this chapter.

1.3. Requirements for high repetition rate

With the light sources of 10–100 PW peak power, the accelerated electron beams can reach the energy up to TeV and ion beams up to GeV, (see Figure 1) as well as by using these secondary sources the ultra-bright X and Y-rays can be obtained [21]. These results could be widely applied into many areas of science, industry, medicine, homeland security, and so on. Nevertheless, it will be possible if the ultra-high peak power laser systems also will be able to combine with high repetition rate (hundreds of Hz to kHz) or high average power (kWs). In the petawatt class laser amplifiers, a pump pulse energy exceeds a few hundred J regime, which means significant thermal load in the gain medium even at low repetition rates.

Thin disk laser technology (TDT) is able to eliminate thermal distortions and damages of the laser crystals in the systems with both high peak and average output power [22]. However, conventionally used in TDT, Nd:YAG and Yb:YAG possess the narrow emission spectra and the low emission cross-section that lead to very complicated multipass amplification schemes which is practically acceptable only for low peak power systems with the ps-level pulse duration. The most promising crystal with required characteristics for ultra-high peak power laser is Ti:Sa, especially if one is taking into account its higher emission cross-section and thermal conductivity (compare 10 W/(mK) for YAG to 40 W/(mK) for sapphire at the room temperature and more than two orders higher when cryogenically cooled). But the attempt to increase the longitudinal gain in TD-amplifier by rising the concentration of active ions and/or using crystals with higher emission cross-section leads to dramatic increase of the gain in the transverse direction and consequently to the big losses and inability to store pump energy due to TASE. Technology for solving this problem will be presented below in this chapter.

2. EXTRACTION DURING PUMPING (EDP) METHOD

As it was mentioned above, the main limitation that arises on the path toward ultra-high output power and intensity of the CPA laser systems is the restriction on the pumping and extraction energy imposed by TASE and TPG within the booster and final large aperture amplifier volume [23].

The reflectivity reduction of the side wall of the gain crystals by grinding, sandblasting and/or coating with an index-matched absorptive polymer or liquid layer in the laser amplifiers is the conventional procedure used to prevent parasitic generation (TPG) [24]. However, the difficulty to find the exact index matching within existing absorbers still restricts the diameter of the pump area to 6–8 cm, corresponding to an extracted energy to around 30 J from Ti:Sa [16]. The amplifier apertures enlarging, or to further pump fluence increasing has led to severe parasitic generation and has failed to increase extracted energy. The additional restriction on storing and extracting energy by TASE in larger gain apertures was demonstrated [25]. TASE necessarily increases with the aperture size, limiting the maximum stored energy that is why this restriction is even stronger than parasitic lasing because the threshold for the latter can be increased due to development of the new index matching materials for absorbers. The uniform luminescence on the left picture of Figure 4 should not delude us, if one will find method to reduce reflections down to zero; the losses remain still incredible big for amplifiers with the large aperture.

The method of the calculation of total volume of TASE radiated out from the crystal during its pumping was developed by Chvykov et al. [25]. Figure 5a shows the evolution of normalized fluorescence of the crystals vs. pumping time when pumped by 100 ns-pulse for different crystal apertures. Here, E_{max} is the theoretical maximum of the extracted energy, and E_{loss} is the lost energy due to TASE. As seen from the plot,

ASE grows dramatically after a certain time of pumping, even for 10 cm—crystal and soon becomes equal to the pumping energy. This means that further pumping is useless because all additional energy will be irradiated out of the crystal as ASE. The critical points of anomalous ASE (APs) are moving to the pumping process beginning with the growing crystal diameter. No more than 20–50% of the pump energy can be stored in the crystals with aperture of 15–20 cm as seen in Figure 5.

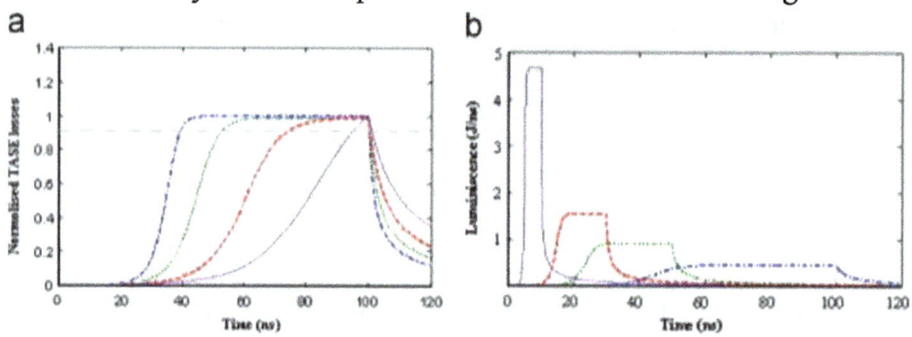

Figure 5. Fluorescence evolution of gain volumes vs. pumping time. (a) Normalized ASE to its maximal value for different crystal apertures, D: blue dash-doted curve D = 20 cm (E_{max} = 824 J, E_{loss} = 590 J), green dots D = 15 cm (E_{max} = 460 J, E_{loss} = 290 J), red dashes D = 10 cm (E_{max} = 210 J, E_{loss} = 100 J), and violet solid D = 6 cm (E_{max} = 74 J, E_{loss} = 20 J); (b) TASE for different pump pulse durations, τ_{pump} (15 cm crystal diameter): violet solid curve τ_{pump} = 10 ns (E_{loss} = 325 J), red dashes τ_{pump} = 30 ns (E_{loss} = 321 J), green dots τ_{pump} = 50 ns (E_{loss} = 315 J), blue dash-dots τ_{pump} = 100 ns (E_{loss} = 290 J).

Shortening of the pump pulse duration does not help to reduce the losses, at least until pulse duration becomes shorter than the time-length of the light distribution through the crystal in transverse direction. This is about several 100 ps, and thus such a pump would be useless due to the very low damage threshold. Fluorescence during pumping for different pump pulse durations are shown for 15 cm crystal diameter (Figure 5b). As seen, the losses slightly grow due to the higher pump rate that enhances TASE with reduction of the pump pulse duration.

If the total compensation of transverse gain by the index-matched absorber coating is not possible, the reflectivity from the side wall will be enough for parasitic generation-TPG. Therefore, one can establish the TPG threshold as an equality transverse gain of the crystal to absorption coefficient of its side wall. There are two most probable transverse modes which can develop: first one is due to the maximal population inversion which is a generation near two working parallel surfaces [23] and the z-pass between these surfaces due to total internal reflection [26] (Figure 6a). The latter mode surpasses the first one in large aperture crystals with a high aspect ratio and respectively low crystal doping because higher gain for second mode. One can define the TPG threshold small signal gain G_t through the pump intensity F_p as:

$$G_t = \exp\left(\frac{6}{zh\nu_p} D F_p \right) \tag{1}$$

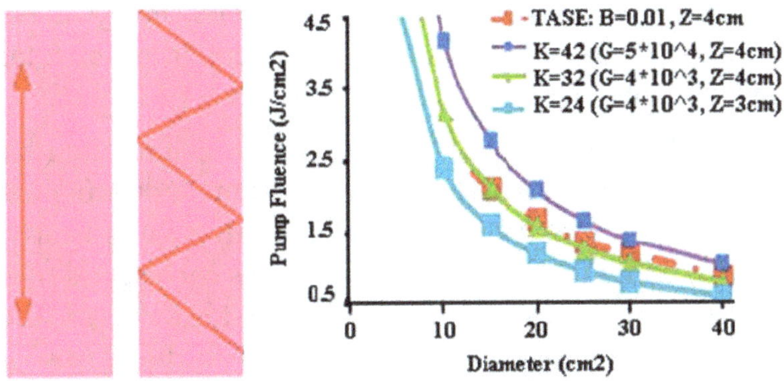

(a) (b)

Figure 6. (a) Two transverse optical paths with highest gain for TASE within a crystal; (b) dependence of pump fluence on crystal diameter: solid curves: calculated for TPG (K = 24, light blue squares; K = 32, green triangles; K = 42, blue squares); dot-dashed curves: calculated for TASE (red squares: B = 0.01).

Introducing $K = D \cdot F_p$ as a K-parameter, we can find the dependence of pump fluence TPG threshold on the crystal diameter (Figure 6b) calculating K for different crystal geometry and used absorber.

$$K_s = \frac{\nu_p}{\nu_{em}} * \ln G_t * \frac{zF_s}{\ln B}; \quad K_z = \frac{\nu_p}{\nu_{em}} * \ln G_t * \frac{zF_s}{2n} \tag{2}$$

There (2) are two formula of the K-factor calculations for two transverse parasitic modes consequently, the generation near working surfaces (K_s) and the z-pass between them (K_z), where D is the pump area diameter, G_t is the highest transverse gain that would be compensated by an index-matched absorptive coating, z is the crystal thickness, n is the index of refraction of the crystal, and the absorption coefficient for the pump frequency—B.

A crystal with a given aperture fixes the maximum pump fluence because the product of crystal diameter and the pump fluence is constant for each value of the critical transverse gain or index matching of the absorber and crystal thickness. Therefore, enlargement of the crystal aperture requires reducing the pump fluence. Moreover, the pump fluence will be limited for any crystals and absorbers because the complete matching is impossible for both polarizations of the transverse modes. The dependence of the maximal possible pump fluence on the crystal diameter was shown in Figure 6b, where the purple rhombus-marked curve is the ideal case of upper bound for 3 cm crystal thickness and reflectivity of σ-polarization under the condition of total index matching for π-polarization.

The same dependences can be built for TASE limitation based on the APs (Figure 5a, crosses with gray dashed line indicating 0.9 of the maximum volume). TASE curve for crystals with B = 0.01 nearly match the TGP curve with K values of 32 (these values correspond to existing absorbers [16]) and is located below the purple ideal curve as shown in the Figure 6. The main conclusion to be drawn from this correspondence is that there remains no motivation to develop better absorber materials, owing to restrictions of TASE. Therefore, the situation for laser amplifiers with large apertures appears bleak, especially with respect to their application as final amplifiers in very high power laser systems. For example, one can conclude from Figure 6, that for 20 cm crystal diameter, the maximum F_p is 1.5 J/cm^2 meaning the extracted fluence is about 0.7 J/cm^2, when saturated one for Ti:Sa crystal is 0.9 J/cm^2, which indicates the deeply inefficient amplifier operation. Nevertheless, below, we will demonstrate that the parasitic losses due to both TASE and TPG can be significantly reduced using EDP technique.

We suggest to change the conventional method of pumping and amplifying the multipass amplifiers to overcome the restrictions discussed above. Conventional method is based on the energy stored in the amplifier media prior to the arrival of the first pass of the input pulse [26]. We are able to forestall TASE and parasitic lasing and increase the extracted energy by continuing to pump after the arrival of the amplified pulse. In this case, the energy extracted during one pass of the amplified pulse through the crystal could be restored by pumping up to AP or TPG threshold before the next pass. An extended pump pulse duration ranging from tens to hundreds of nanoseconds, or several delayed pulses is required for EDP process. One can get sufficient time in this case for proper pumping between passes, allowing increased total pump fluence due to the longer pump pulse duration and overcoming problems with temporal jitter. This approach was shown to double the output flux above the parasitic lasing limit in the experiments in University of Michigan [27].

Optimization of EDP method for presently available large aperture Ti:Sa amplifiers was made in Ref. [28] . The description of amplified transmission of pulses through the multipass amplifiers can be done by broadly applying Frantz-Nodvik solution for the 1-D photon transport Eq. [26]. Solution for optimal EDP-amplifiers can be rewritten as a single equation, in contrast to conventional amplifiers, where this equation is applied iteratively with adjustment of small signal gain for each pass, because the restoration of the population inversion by the pump between passes in case of EDP, and hence of the small signal gain, for each pass:

$$F_{out} = F_s \ln\left\{1 + \left[\exp\left(\frac{F_{in}}{F_s}\right) - 1\right] \exp\left(N \frac{\nu_{em}}{\nu_p} \frac{F_p}{F_s}\right)\right\} \tag{3}$$

where N is the number of passes, ν_{em} is the emission frequency, ν_p is the pump frequency, F_{out} is the N-pass output fluence, F_s is the saturation fluence, F_{in} is the incident fluence, and F_p is the initial pump fluence (before signal arrives).

Therefore, we simply introduce a factor of N into the small signal gain expression to get the output flux after N-passes of amplification. This could be easily proved by introducing the output flux from any N-1 passes as F_{in} for N-th pass and following the usual iterative procedure.

Dependences of the output fluence calculations for 4-pass amplification with conventional amplification and EDP for various pump fluence are presented in Figure 7a. From this graph, one can get the next conclusions. First—EDP amplifiers can deliver significantly more energy (up to four times for four-pass amplifier) compared to regular ones with the same initial pump fluence. Second—the input fluence that saturates the amplifier is much higher than that for the regular case because the graphs asymptotically approach to the value of four times of the initial pump flux multiplied by coefficient of quantum defect. From comparison of the two green or two blue curves in Figure 7a, one can see that the curvature slope becomes twice lower at 50–70 mJ/cm^2 of the input fluence for conventional amplifier and near 200 mJ/cm^2 for EDP-amplifier. We have a higher extracted energy with growing input flux and so are able to add more pump fluence between passes. So, the input fluence for amplification with EDP has to be much higher to make them efficient and the process of amplification mimics the case of a high value of F_s. On the other hand, this says us about a much higher energy capacity of the amplifier.

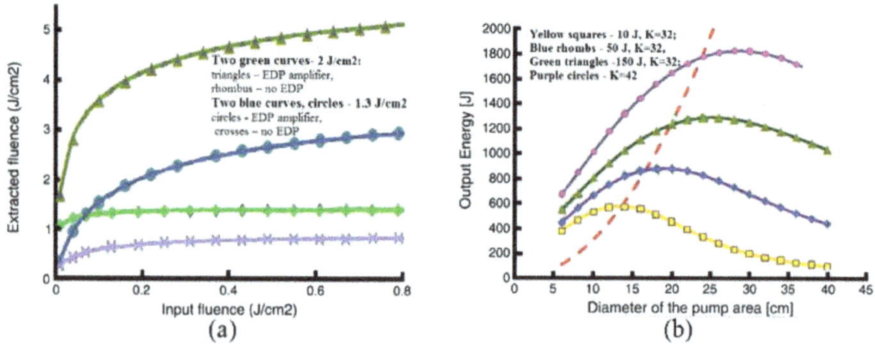

Figure 7. (a) Dependence of the output fluence of the 4-pass amplifier on the input fluence for several pump ones; (b) dependence of the output energy on diameter of pumped area for different input energies and K-factor; dashed red curve: damage threshold fluence.

The optimal diameter for the pump area of the EDP-amplifiers can be demonstrated if into Eq. (3), in place of F_{in} and F_p, will be introducing the ratios E_{in}/A and $K/D(2)$, respectively, where E_{in} is the total incident energy and A is the pump area. The dependences of the output energy on diameter of the pump area, for different K values and incident energies, are presented in Figure 7b. The output fluence of damage threshold was marked by the red dashed line. Using these graphs, we can calculate the highest output energy for amplification with EDP for a practically available liquid absorber and a crystal with 20–25 cm diameter [29]. This absorber is able to accommodate the maximum transverse gain of 4000, which leads to a value of $K = 32$. We can find the output energy of ~800 J from the blue rhombus curve in Figure 7b, which corresponds to 50 J input energy. In Ref. [16], the authors demonstrate the final EDP-amplifier 60 J, making 800 J realistic output energy for the next stage of amplification. This output energy can be reasonable for crystals with this diameter due to output fluence 2 J/cm^2 and the extraction efficiency close to theoretical one 65% which is reachable for EDP-amplifiers.

The EDP-technology was successfully spread out in the many world class laboratory for application in the ultra-high power laser systems, demonstrated today's world record near 200 J extracted energy and 5.0 PW output power [30]. Below several examples will be exhibited.

First time, this method was used in the 4-pass amplifier of the HERCULES-300 TW Laser [31]. Four-pass amplifier crystal was cooled cryogenically to 120 K avoiding wave-front thermodistortion of the output beam. Because the amplifier crystal was located in the vacuum chamber to avoid the surface deposition, cladding of the side surface was impossible. Extracted flux was increased from 0.6 to 1.2 J/cm^2 due to application of EDP [27].

The authors of [16] have successfully developed a high-energy Ti:Sa laser system that delivers 33 J before compression at 0.1 Hz using LASERIX-4-pass EDP-amplifier. 100 mm diameter Ti:Sa crystal (pumped area-60 mm) was used as the final amplifier which was pumped with 72 J of energy delivered by frequency-doubled high-repetition rate Nd:Glass lasers. The good amplification efficiency of 45% with a homogeneous flat-top spatial amplified intensity profile was finally obtained.

The EDP-method has also been applied on several Ti:Sa booster amplifiers of petawatt scale. One of them was 3-pass EDP-amplifier of APRI 1.5 PW CPA-laser (South Korea) [32]. The team of laser developers reported about the generation of 1.5 PW by using two stage final EDP-amplification with the maximum output energy of 60.2 J at a pump energy of 120 J.

Shanghai Institute of Optics and Fine Mechanics, Chinese 5 PW CPA: laser, 4-pass EDP-amplifier [30]. Effective suppression of the parasitic lasing in the final booster amplifier was done using the EDP-technology combined with index-matching cladding technique and the precise control of the time delay between the input seed pulse of 35 J and pump pulses of 312 J. The output energy of 192.3 J from the final amplifier was corresponding to a pump conversion efficiency of 62% to the output laser energy.

The design of the final EDP-amplifiers was recently developed for three pillars of the Extreme Light Infrastructure (ELI) project [33]. ELI is the ambitious pan-European laser research project. The major mission of the ELI facility is to make a wide range of cutting-edge ultrafast light sources available to the international scientific community. The first purpose of the facilities is to design, develop and build ultra-high-power lasers with focusable intensities and average powers reaching far beyond the existing laser systems. The secondary purpose is to contribute to the scientific and technological development toward generating 200 PW pulses, being the ultimate goal of the ELI project. PW-class lasers have been planned to build in the three pillars of ELI. 2 PW peak power, 10 Hz repetition rates and <20 fs pulse duration lasers will be part of the ELI-BEAMLINES and the ELI-ALPS, while the L4 of the ELI-BEAMLINES as well as the ELI-NP lasers aiming at 300 J/10 PW lasers. The roadmap for 200 PW laser facility was paved by the ELI consortium. This will increase the available laser power by at least one order of magnitude in its first three pillars, and on another more order of magnitude in its fourth ultra-high-intensity pillar. The laser power frontier was planned to be pushed into sub-exawatt regime by the establishment of ELI s fourth pillar.

The examples of final EDP-amplifiers design of 10s PW laser system are presented on the Figures 8 and 9. As the estimations demonstrate, the EDP-technology is able to significantly increase the output energy and intensity. There are calculated as optimal output parameters: the diameter of the pump area and crystal ~19 and 20 cm, the pump energy ~960 J, the input and the output energy ~60 and 600 J. The losses with the EDP technology can be made under 5% (Figure 8).

The total losses in 4 cm thickness crystal of conventional amplifier is about 70%, and for 3 cm, it is ~80% as seen from the Figure 8a, and significant gross begins from 30 and 20 ns of the pump, respectively. On the Figure 8b, dependences of losses for optimal extraction of the 3-pass and 4-pass EDP amplifier are demonstrated with the crystal thickness of 4 cm.

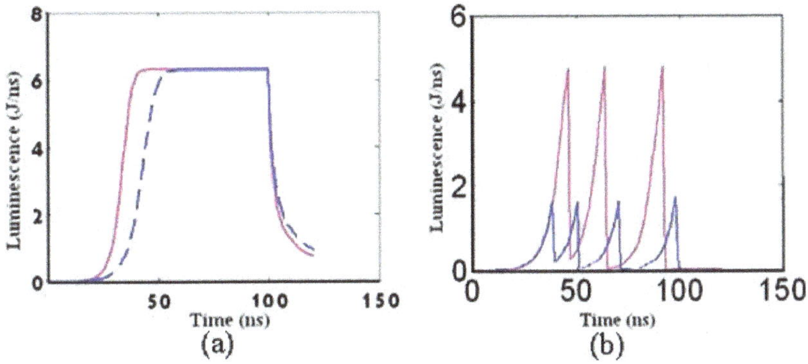

Figure 8. Dependences of losses for 4 (blue-dashed curve) and 3 cm (pink-solid curve) thickness of the crystal: (a) conventional amplifier; (b) EDP-amplifier.

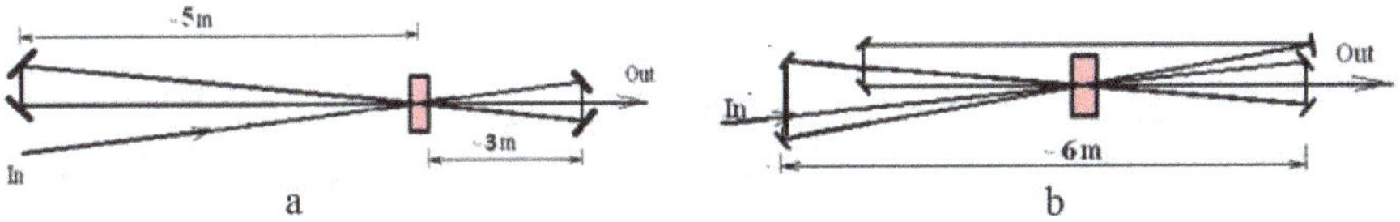

Figure 9. Final EDP-amplifiers with 3- and 4-passes: (a) EDP amplifier with a 19-cm-diameter pump area can approach 500-J energy with a 60-J input (100 J losses) during 3-pass; (b) 4-pass EDP amplifier with a 19-cm-diameter pump area can approach 600-J energy with a 60-J input, when pumped by 960 kJ (28 J losses).

For 3-pass, the delay between passes 1 and 2 is about 20 ns and 2 and 3–30 ns, whereas the scheme with delays between passes of the 4-pass amplifier (15, 20, 30 ns) is presented also in the Figure 8b. Taking in account the compressor transmission efficiency 70% [32], the compressed energy up to 400 J can be expected. Output peak power about 30 PW can be reached in one channel with pulse duration 10–15 fs. This amplifier could serve as a building block for ELI fourth pillar, and seven EDPCPA channels will be enough for approaching 200 PW.

As a final remark of this part, we can emphasize that there is a big gap between the reached now output energy of 200 J and potential possibilities of the EDP amplifiers of 600–800 J extracted energy with today's available Ti:Sa crystals, which should be filled in the closest future. Moreover, manufacturers are working now under the larger aperture Ti:Sa crystal and have reached 30 cm crystal diameter. The estimations demonstrate the 40-cm crystals with 6-cm thickness which is nearly maximum for EDP amplification due to geometric reasons.

This amplifier is able to supply ~3 kJ/140 PW with 250 J of seed energy while suffering TASE losses of about 207 J. Two EDPCPA channels will be enough for approaching 200 PW in this case.

3. PULSE DURATION SHORTENING

3.1. Preserving and restoring pulse spectral bandwidth

Here, we will concern another way of the peak power increasing, namely pulse shortening. It was discussed above that the gain narrowing and saturation limit the achievable pulse duration to about 30 fs [30, 31]. To achieve the 10–15 fs pulse duration designed for the Apollon system as well as for the 10 PW laser of the ELI [18], further scientific and technological efforts have to be made to avoid spectral narrowing in the power amplifiers.

The idea to use optical rotatory dispersion (ORD) (the angle of polarization rotation dependence on wavelength) spectral filter was suggested for conservation of bandwidth in low gain multipass Ti:Sa amplifiers, typically used for the intermediate and duty end amplifiers of multi-TW-PW class systems [34]. The spectral gain can be effectively re-shaped using difference of π- and σ-emission cross-sections of Ti:Sa crystal which will result in the keeping bandwidth of efficient amplification [13]. The overall amplification process can be kept lossless and efficient as the effective spectral gain is tuned, unlike to the spectral shaping of the pulse. This technique is especially advantageous for amplifiers where the energy extraction is of major importance in intermediate and duty end ones of ultrashort pulse large-scale laser systems.

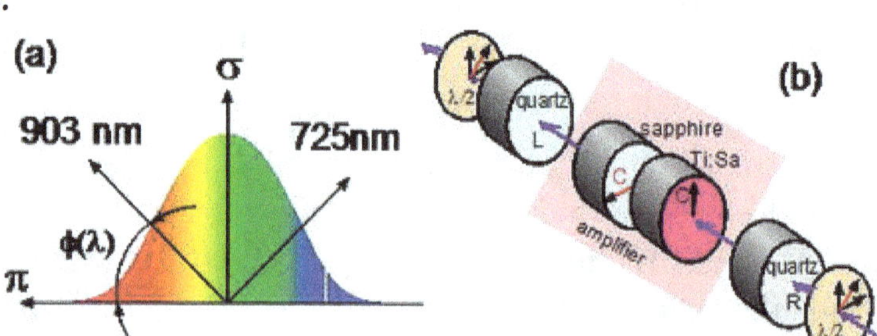

Figure 10. Distribution of polarization directions of spectral components by a 17.4 nm ORD quartz (a); principle schematic of the polarization encoded amplifier (b).

The optical rotatory dispersion (ORD) is applied to encode/decode the polarization state of the amplified spectrum in addition to using both π- and σ-emission cross-sections. Polarization vectors of spectral components are encoded before amplification using an ORD quartz crystal (right rotating quartz in Figure 10b) with a resulting distribution between π- and σ-cross-sections (Figure 10a).

The emission cross-section of σ-polarized light is nearly 0.4 of the π-polarized light. The sides of the spectral band are aligned toward the π-axis, while the components around the central and most intense part of the spectrum are directed closer to the σ-axis. This allows shaping of the spectral gain so the gain narrowing can be substantially reduced or eliminated at all. The second quartz crystal (left rotating quartz in Figure 10b), with an opposite sign of ORD decodes the polarization state of spectral components after amplification. The whole polarization encoded spectral distribution fits between π- and σ-directions by using of two achromatic λ/2 plates. A temporal walk-off between the π- and σ-components due to birefringence in Ti:Sa can be compensated by the addition of an undoped, but orthogonally oriented sapphire of the same thickness.

The 10 Hz, mJ laser system of the Max Born Institute was using proof-of-principle experiment (Figure 11). The 20 fs pulses from oscillator were stretched to 120 ps and amplified in a 10-pass Ti:Sa amplifier to ~1 mJ. The pulses with 35 nm bandwidth were compressed using a diffraction grating compressor to their transform limited pulse duration of 28 fs.

The computer modeling of the PE-CPA amplifier resulting in the broadest bandwidth showed an optimum thickness of 17.5 mm of the quartz crystal (Figure 10b). A spectrally dependent model of amplification of strongly chirped pulses was used for modeling the PE-CPA amplification. It includes ORD encoding, saturation of amplification and decoding of polarization vectors. The polarization encoded amplifier was built in a 6-pass configuration with a Ti:Sa crystal of 15 mm thickness (Figure 11). It was pumped from both sides by the second harmonic beam of a Q-Switched Nd:YAG laser. The pump diameter was de-magnified from 6 mm to 2.5 mm by two lenses (focal lengths 1 m) to fit it with the diameter of the seed pulse. An undoped sapphire was mounted next to the Ti:Sa crystal (both with thickness of 15 mm). Both crystals had orthogonally directed C-axis to compensate for the inherent temporal walk-off. The incident angles of amplified beams on the active medium were kept as small as possible to minimize uncompensated walk-off and the total length of the amplifier was set to 2.1 m. To eliminate the spectral interference fringes from the outgoing amplified spectrum, the sapphire crystal was mounted on a fine rotator stage and was adjusted. The compressed pulses were loosely focused into a 5 mm BK7 glass plate by a lens (focal length 3 m) with the most homogeneous part of the SPM signal filtered out by a pinhole located at the focus of the lens. Using this technique, the bandwidth of the seed pulse was broadened in self-phase modulation (SPM) stage and stretched in the bulk (Figure 11) to show the broadband amplification capability of the scheme.

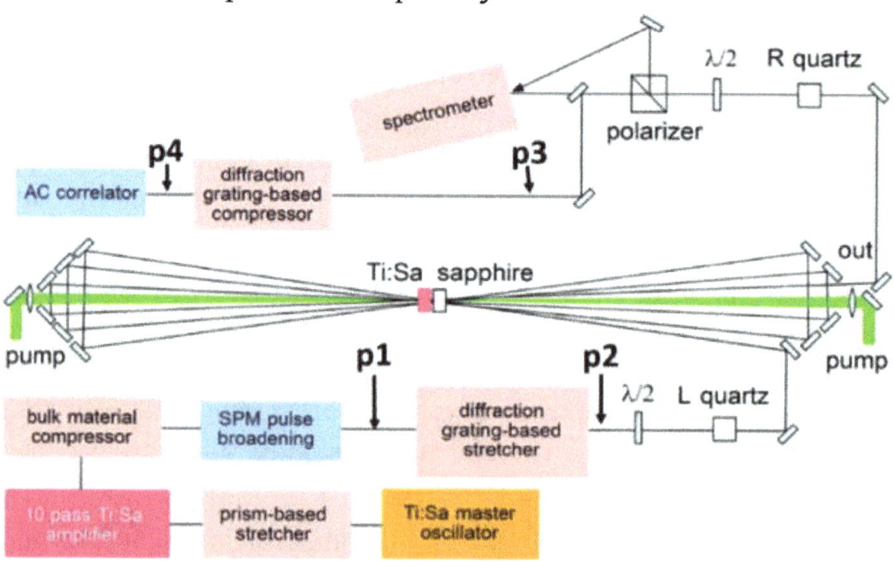

Figure 11. Schematic of the experiment system. Spectra were measured at places p1–p4.

A multi-passing SF6 glass block (80 cm path) was used for stretching the broadband pulse with a near top-hat spectral profile and a bandwidth of 91 nm, the last one was then collimated by a lens (focal length 1.5 m). The seed pulse energy was about 35 μJ due to the energy losses at the pinhole, as well as two metallic mirrors used in the bulk material stretcher and reflections on surfaces of the uncoated ORD quartz.

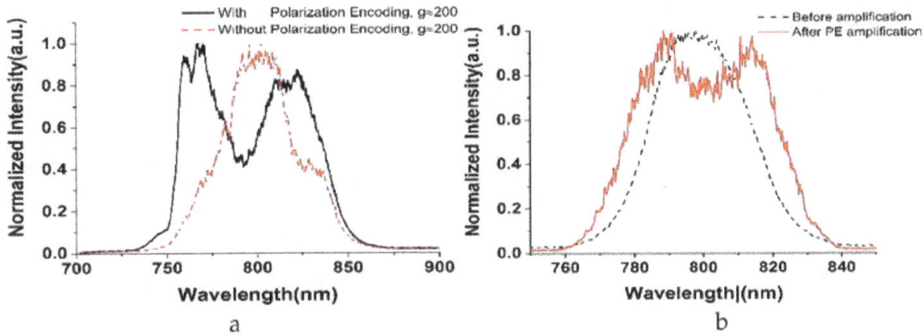

Figure 12. (a) Amplified spectra with (solid line) and without (dash line) the PE technique under a high gain amplification; (b) spectra before (dash line) and after the PE amplifier (solid line) at a gain of 30.

The polarization encoded amplification was tested for a high (~200) and low (~30) gain case to demonstrate the effectiveness of the PE-CPA method. The broadband seed was initially amplified without polarization encoding in order to form a baseline (Figure 12a) with a gain of about 200. The gain-narrowed feature with a resulting bandwidth of less than the half of the seed was demonstrated at the recorded spectrum. The output spectra closed to the input seed resulted in FWHM bandwidth of 82 nm using polarization encoded amplification. A dip was produced at the center due to the spectral edges of the spectrum that experienced a high gain. The following experiments demonstrated that spectral-dependent gain feature can even broaden the bandwidth depending on the initial shape of the seed spectrum and an amplifier saturation.

Seed pulses (energy ~200 μJ) were sent directly to the amplifier, by-passing the compressor and SPM stages to investigate the low gain scenario. A pulse with a bandwidth of about 48 nm was achieved after polarization encoded amplification with a net gain of ~30. In contrast to the top-hat seed spectrum (Figure 12a), the amplified Gaussian spectrum is almost 40% broader compared to the seed due to the high-gain experienced by the spectral wings (Figure 12b).

It was also shown that an additional polarization rotation takes place during the pulse amplification, substantially changing the conditions of high-energy amplification and introducing the distortion into initial polarization encoding. Despite the additional polarization rotation occurring during amplification impeding the decoding process, the energy efficiency of the back conversion to the linear polarized state can be improved by increasing the thickness of the second quartz crystal. The computer modeling shows that an efficiency as high as 99% to convert energy to a linear polarized state for moderate gain values can be reached with the optimal thickness of the second quartz.

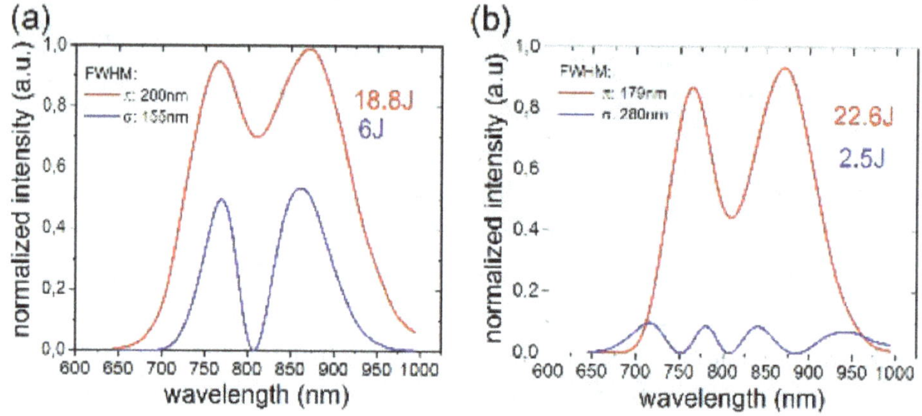

Figure 13. PE-CPA amplifier (simulation): (a) thicknesses of encoding and decoding quartz plates are equal (17.4 mm); and (b) the second quartz plate is 35 mm thick. The seed pulse has a Gaussian spectrum with FWHN = 180 nm. Pump energy: 50 J, seed energy: 2 J, 4 passes.

Good compressibility of spectrally shaped PE amplified pulses is highly important for the practical application of the PE-CPA method. For testing its compressibility, the pulse spectrum was broadened to ~72 nm during the PE amplification and then compressed down to compressor consisting of two diffraction gratings. The compressor reduced the bandwidth to ~ 60 nm. Single-shot second-order autocorrelator (AC) was used for pulse duration measurements. The width of the measured AC trace is 28.0 fs which corresponds to the pulse duration close to 18.5 fs.

The computer modeling shows that a 200 nm spectrum is achievable at a multi-joule level in a PE-CPA Ti:Sa amplifier with seed pulses broader than the experimental bandwidth. This output requires seed pulses with a smooth spectrum preferably with a Gaussian shape. The computer modeling of the output spectras of 4-pass PE-CPA amplifier seeded by a 2 J Gaussian-shaped pulse (FWHM of 180 nm) and pumped by 50 J is shown in Figure 13. The spectrum is broadened to 200 nm after amplification. Nearly, 30% depolarization losses (Figure 13a) can be compensated by doubling the thickness of the decoding quartz (Figure 13b).

High-energy polarization encoded Ti:Sa amplifiers are able to deliver an amplified bandwidth of 200 nm, which was predicted by detailed modeling, making it a promising technique for intermediate and final amplifiers of high field Ti:Sa CPA-laser systems. This technique may pave the way to PW class Ti:Sa lasers with tens of Joule few cycle laser pulses.

3.2. Multiple compression stages

The multiple compression method based on spectral broadening using self-phase modulation (SPM) in the bulk of material with the further recompression of the chirped pulse is able to deliver even shorter pulse duration (potentially below 10 fs) without energy sacrificing. Previous attempts were not suited for the highest powers because it used SPM with combination of spatial filtering in near field [35, 36] in order to provide spatially uniform phase due to the axial uniform part of the Gaussian distributed intensity. These approaches were restricted to only mJ-level of the output energy.

In Ref. [37], the next logical step was suggested to use super-Gaussian beam profile due to its higher uniformity. If a laser beam has spatially uniform intensity, SPM can be used in near-field without spatial filtering and consequently eliminating constraints of the laser energy and transmission efficiency. In a proof-of-principal experiment, spectral broadening of 30 nm pulse spectra to ~80 nm was demonstrated. A flat-top beam profile of the HERCULES laser [31] at 30TW power level was used for SPM spectral broadening and compression of a fraction of this power from 30 to 14 fs by a prism-based compressor was fulfilled. The laser pulse with the energy about 1 J and 6 cm beam diameter was directed at intensity ~1 TW/cm^2 on the bulk of fused silica plate after compression by the standard grating compressor to 30 fs. Glass plates of 0.8, 2 and 2.5 cm thicknesses were used for SFM in the experiments. After that, the 100 times attenuated energy and reduced to 1 cm diameter beam was directed into the prism compressor. The spatially resolved spectrum of the pulse was measured and shows the deviation of the energy within 6% through most part of the beam aperture. It follows from Figure 14a that the initial FWHM of the space-integrated spectrum ~30 nm was broadened to ~80 nm with the 2 cm glass plate.

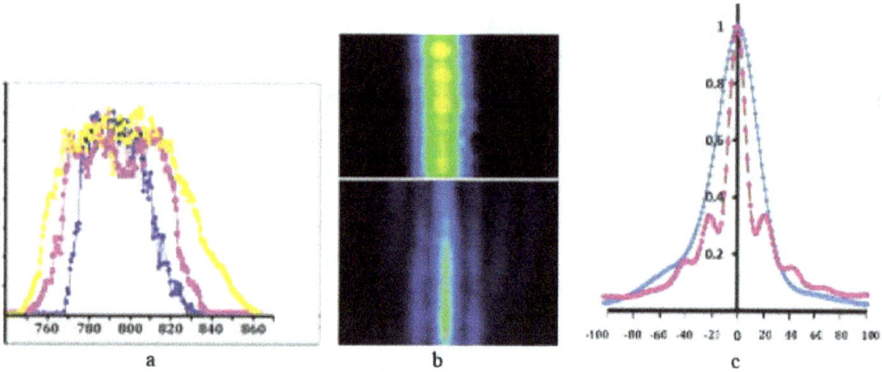

Figure 14. (a) Pulse spectrums: yellow curve (triangles): SPM with glass plate of 2 cm thickness, pink curve (squares): with plate of the 0.8 cm and blue curve (diamonds) represent the spectrum without the glass plate (no SPM); (b) autocorrelation of compressed pulse after SPM, (c) autocorrelation trace of the initial transform limited pulse (solid blue line) and compressed pulse after SPM (dashed pink line).

The final compressor consisted of two prisms (glass SF10) separated by ~50 cm, after which the single-shot autocorrelator was used for pulse duration measurement. Figure 14b and c demonstrate the autocorrelation picture and trace lineout, respectively. The initial nearly-transform limited pulse, with the duration from the FWHM autocorrelation trace of ~30 fs was shown on Figure 14b and the solid blue curve in plot (c) inferred assuming Gaussian pulse-shape. The pulse after the SPM used with the 2.5 cm glass plate and recompression is shown by autocorrelation (b) and the dashed pink curve in plot (c). The compressed pulse duration exhibits shortening by a factor of 2 leading to ~14 ±1 fs.

During these experiments, the instability from shot to shot of the output pulse shape, distribution of the significant part of the energy into wings (Figure 14b and c) and high level of the white light generation were observed. The distortion of the output pulse shape and its instability is a result of uncompensated third-order dispersion (TOD) of prism compressor that was used as second stage of compression. Besides, TOD coefficient in nonlinear Schrödinger equation of laser pulse propagation through transparent media is inversely proportional to the third power of the pulse duration and the pulse duration consequently varies inversely with incident peak intensity due to SPM. This is why even small fluctuation of the incident intensity leads to significant distortion of the shape of the compressed pulse. A solution of these two complications could be the replacement of the prism compressor with chirped mirrors with compensated TOD. Using about 10 bounces on the chirped mirrors and the glass plate at Brewster angle promises compressor throughput above 90%.

On the other hand, reducing the white light generation could also be possible since the main processes involved in this generation are efficient only when phase matching conditions are fulfilled, which could be destroyed by correctly choosing incident intensity and glass plate thickness. Evidence of such possibility can be found in Ref. [36], where the SPM was done with a 3 mm glass plate and incident intensity of 8 TW/cm^2, and resulted in a very low level of the white light generation.

Further development of this idea was done in Ref. [38], where SPM in thin film (within 1 mm) and much higher intensity (up to tens of TW/cm^2) was suggested. This technique is able to significantly reduce the instability and white light generation discussed above and is capable to compress 25 fs large energy pulses as high as 1 kJ to the 1–2 fs level. In order to demonstrate the potential of thin-film-compressor, numerical simulation was done with the following initial beam parameters: central wave length 800 nm, pulse duration T = 27 fs, flattop transverse intensity distribution with diameter 160 mm and total energy 27 Joules. Thicknesses of the first and second thin film elements for SPM are 0.5 mm and 0.1 mm. The fundamental peak intensity is 4.7 TW/cm^2, after the first and second stages with temporal recompression procedure 16.6 TW/cm^2 and 43 TW/cm^2 at pulse durations 6.4 fs and 2.1 fs correspondingly.

As a conclusion of this part of chapter one can deduce, the discussed technologies of the pulse shortening are able to reduce the pulse duration to additional one order without energy sacrificing, which means the output peak power of a few hundred PW are possible from the single channel of CPA laser systems.

4. HIGH REPETITION RATE OF ULTRA-HIGH PEAK POWER LASER SYSTEMS

As it was discusse, the applications of the ultra-high peak power laser pulses required also high repetition rate or the other words high average power laser systems. Limitation of the high peak power laser systems on the repetition rate can be overcome by TD-technology and hence can increase their average power. Making the active medium thin reduces the longitudinal gain, which can be compensated by increasing the concentration of active ions and/or using crystals with higher emission cross-section, and the most promising crystal with required characteristics for these amplifiers is Ti:Sa. However, the attempt to increase the longitudinal gain leads to a dramatic growth of the gain in the transverse direction of the active medium. Therefore, TASE and TPG are strongly dependent on the axial gain and the ratio of the pumped area diameter to the crystal thickness (aspect ratio). For thin Ti:Sa active media, these effects cause significant depletion of the inverted population, and hence limit the extracted energy, the same way as it was detailed discussed in Section 2.

The EDP-technique was suggested to be applied for TD crystals [39] and successfully tested [41] mostly in powerful final stages of ultra-high peak power laser systems with hundreds of TWs to tens of PWs power, which are operating in the saturation regime. The overall gain of the amplifier can be kept as small as ~10–20, since the extraction of energy is a major importance. On the other hand, a large amplifier aperture is required for the high energy level. Therefore, power stages of the said Ti:Sa systems require a crystal size ranging from 5 to 20 cm, contrary to "conventional" TD amplifiers with the aperture of few millimeters where several tens of passes can be done [40]. The reasonable number of passes through the amplifier in this case, is restricted up to 6 due to geometrical complexity and available space. These two requirements form the lower boundary for the small signal gain of 3–5 per pass or the saturated one of 1.5–2.

Figure 15a demonstrates dependence of the transverse gain on the crystal aspect ratio for different longitudinal small signal gain values, which correspond consequently to the absorbed pump fluence F of G_l = 10 – F = 3.8 J/cm^2, G_l = 7 – F = 3 J/cm^2, G_l = 5 – F = 2.5 J/cm^2 and G_l = 3.5 – F = 2 J/cm^2. The maximal possible suppression of transverse gain using a conventional method of the side surface cladding and its combination with the EDP-method correspond two horizontal lines (solid and dashed line in the figure). The side surface cladding with a commonly used liquid absorber is considered in both cases. The conventional method of transverse gain compensation supports the aspect ratio between 2 and 4, as seen from this picture, while the EDP-amplifiers can afford 8–15. That means the latter one can be applied as TD-amplifier and can increase the output energy up to 16 times. The maximum value of aspect ratio can be determined from these calculations taking into account the required output energy and crystal thickness. Then, the small signal gain could be chosen depending on amount of pump passes and the crystal doping that will support the required absorption. For example, EDP is able to afford for the small signal gain of 3.5, the highest aspect ratio 15 with the reasonable amount of signal passes of 4–5. So, the Ti:Sa crystal of 15 cm is required for 10 PW laser (300 J); this corresponds to the crystal thickness of 1 cm and doping according to the chosen pump absorption.

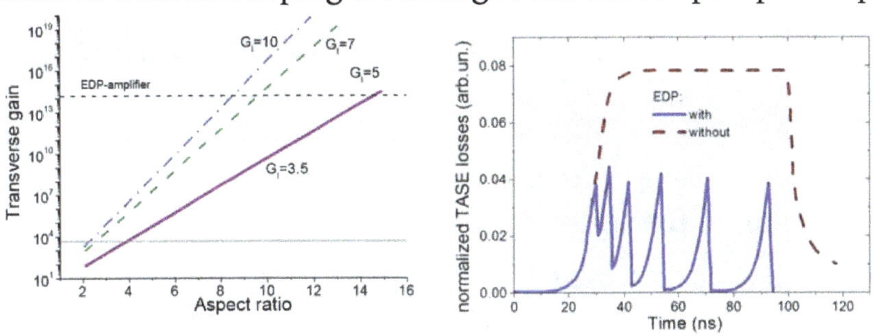

a b

Figure 15. (a) Dependences of transverse gain on crystal aspect ratios; (b) losses calculated for the 200 TW/100 Hz 6-pass Ti:Sa EDP-power amplifier. Shaded area is the pump pulse.

As for conventional amplifiers, the EDP method can be applied in a similar way to thin disks amplifiers [28]. This method allows in optimum conditions to significantly reduce the losses in the crystals with the big aspect ratio, or in thin disk crystals (to 5–15%). The EDP method requires an extended pump-pulse duration ranging from tens to hundreds of nanoseconds, or a train consisting of several delayed shorter pulses. EDP can then be naturally combined with thin disk Ti:Sa amplifiers with regularly doping crystals because a smaller portion of the pump energy can be absorbed in smaller crystal thickness per pass. Choosing the correct distances of the pump and the seed pass shoulders, the multipassing pump can be adjusted for the optimum EDP. We demonstrate further that a new line of the CPA ultra-high intense high average power laser systems can be opened by combination of EDP and TDT with a possibility to be scaled up to tens of a PW peak power and hundred Hz-repetition rate.

As an example, losses have been calculated in a 20 × 2 mm Ti:Sa crystal pumped with 11 J (532 nm, 100 ns) and seeded by 180 mJ using the method elaborated in Ref. [25] with some modification of the computer model. According to these calculations, TASE loss without EDP (dashed curve in Figure 15b) is ~80%, the output energy for an EDP-amplifier is 6 J, while energy loss (solid curve) is ~10–15%. The peak power of the laser can reach 120–400TW if compressor transmission efficiency is 70% and pulse duration is 10–30 fs, while the average power is close to 0.6 kW at the repetition rate of 100 Hz.

Proof-of-principle experiment of the operation of a broadband EDP-TD amplifier in a 100 TW CPA laser system was presented in Ref. [41]. The amplifier head was fully characterized, including measurement and modeling of the temperature distribution, dynamics of amplification, and wave front of the amplified pulses.

Test bed was assembled in Max-Born-Institute on the base of commercial CPA laser system amplifier, which produces 100 TW peak power, 28 fs laser pulses at 10 Hz repetition rate. The final cryogenically cooled Ti:Sa amplifier has been replaced in experiments with the EDP-TD room temperature cooled arrangement (Figure 16). Ti:Sa crystal with a 35 mm diameter, 3 mm thickness and an absorption coefficient of 2 cm^{-1} was mounted in the homemade thin disk head module. The active mirror scheme was applied for the amplifier. Rear side of the crystal was HR coated, while the front one was AR coated, for both pump (532 nm) and seed (800 nm) wavelengths. The sides surface of the Ti:Sa crystal was coated by the absorbing material (refraction index of 1.76 at 800 nm), for TASE and TPG suppression. The rear surface of the crystal was actively water cooled to room temperature. Three temporally synchronized lasers, each providing 15 ns, 2 J pulses at 532 nm wavelength pumped area of 24 mm diameter.

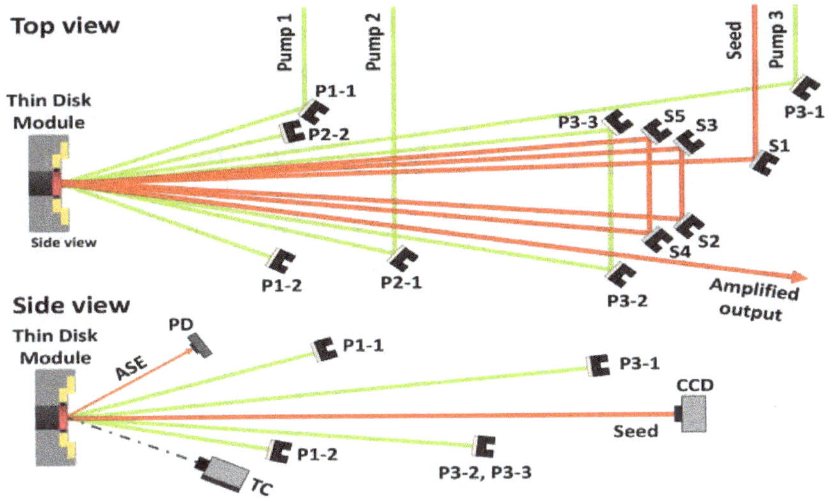

Figure 16. Experimental set up. The layout of the seed amplification consists of mirrors S1–S5 (red lines). The green beams are pump. Mirrors P1-1 and P1-2 were used for pump laser 1, mirrors P2-1 and P2-2 were used for pump 2, mirrors P3-1, P3-2 and P3-3 were used for pump 3.

Each of the three pump beams passed through the thin disk twice, one pass includes the reflection from the rear surface of the crystal. The total absorption for double pass of pump was measured to be 85%. The vertically separated pump beams one can see on the side view of the **Figure 16** with the smallest incidence angles, where the pump 2 was omitted. Three passes (mirrors S1-S5) amplified a positively stretched seed of few 100 ps pulse duration (energy -0.5 J) with a total gain of about 5.

Severe parasitic lasing was generated despite the use of the liquid absorber in the absence of the seed, when the active medium was simultaneously pumped with total absorbed energy 3.4 J by the double passed pump beams 1 and 3. This is clearly visible on the oscillogram of the luminescence (Figure 17a) from Ti:Sa crystal. EDP technique was applied to avoid TPG losses, extracting energy from the crystal before the second pass of the pump 3 (~3 J of absorbed energy). After first pass of the pump 3, the second one added 0.4 J to the stored energy delaying about 20 ns after the extraction of 0.5 J by the first seed pass (total amplified energy after the first pass was about 1 J). To avoid another round of parasitic generation, the pump pulse from the pump laser 2 was added between the second and third seed passes (Figure 17b), which allowed to reach 2.6 J of the amplified energy when the total absorbed pump was about 5 J. Three passes amplification only and two passes of the pump were required for this seed to achieve almost 50% extraction efficiency (compared with tens of passes of conventional Nd:YAG or Yb:YAG TD amplifiers) due to the much greater emission cross-section and thickness of the Ti:Sa crystal. The typical near field cross-section of pump and seed amplified after the third pass beams had the flat-top shape with the good uniformity.

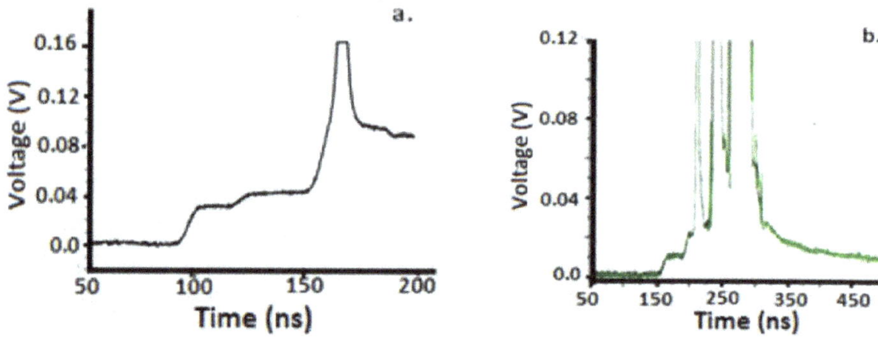

Figure 17. Oscillogram of the luminescence from Ti:Sa crystal. (a) Parasitic generation after double pass of the pump beams 1 and 3 with total absorbed energy about 3.4 J; (b) the luminescence with three seed passes.

The temperature distribution over the crystal input surface in the transverse direction through the crystal was measured (Figure 18a) with a thermal imager. The crystal was pumped with 4 J of pulse energy at 10 Hz repetition rate without the energy extraction by the seed. Average flow speed was about 7 cm^3/s with the initial temperature of the coolant 18°C.

Figure 18b shows the temperature growth dynamic taken at the central point of the crystal until temperature stabilization. As seen from this curvature, the stabilization was reached after the crystal pumping within 20 s with 10 Hz repetition rate of the pump pulses. Figure 18 shows that a coolant temperature of 18°C was sufficient to keep the maximal crystal temperature after stabilization under 30°C. The variation through the pump area was only 3°C, which does not significantly affect the seed beam wave front. The liquid absorber consumed a TASE absorption which is the largest portion of total luminescence; so, most of the pump energy was transmitted to the crystal heating. That means, similar result could be seen with the same crystal pumped with 8 J at the same repetition rate, when half of this energy is being extracted by the seed. Furthermore, a higher coolant flow speed could reduce significantly highest temperature.

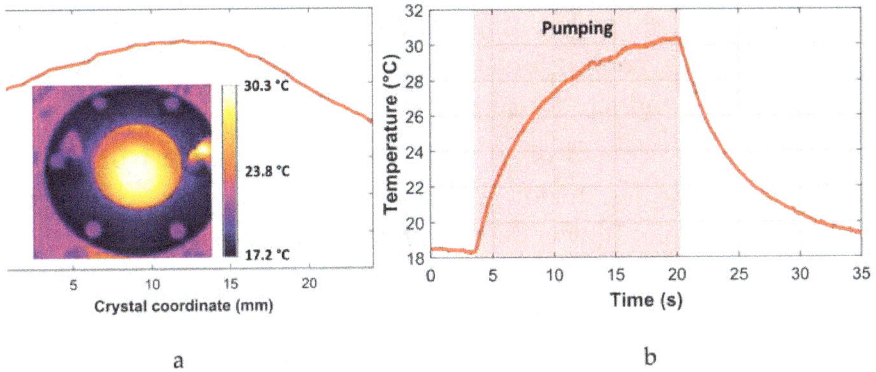

Figure 18. (a) Cross section of temperature distribution in the amplifier crystal after heating stabilization during pumping by the 4 J per pulse at 10 Hz repetition rate. The curve shows the temperature distribution of the horizontal mean section of the pumped area (red, solid line) and temperature distribution in the amplifier head (inset); (b) central point temperature growth dynamic of Ti:Sa crystal.

The impact on the beam wave front was also measured during these crystal heating conditions. Measurements were taken until the crystal temperature was stabilized. The wave front before and after pumping was measured and consequently demonstrated P-V: 0.32 μm, RMS: 0.06 μm and after heating stabilization P-V: 1.51 μm RMS: 0.39 μm. These results are better than the comparable single-shot laser systems with side extraction of the heat measured before wave front correction [42].

Heat extraction from the Ti:Sa thin disk was numerically modeled to match experimental conditions and scale-up the system for higher repetition rates of 100 Hz, corresponding to 360 W of the thermal load. These model parameters account for the 8 J pumping with an energy extraction efficiency by amplification of the seed pulses of 50%. Proper cooling conditions were considered using a flow velocity of 5 m/s, and coolant temperature of 5°C. According to these simulations, the temperature profile is much more symmetric and smoother in the thinner crystal than in thick one. For 3 and 2 mm crystals, the temperature difference between the center and the edges was improved with central peak temperatures of 83.5 and 73.5°C, respectively. The obtained results demonstrate the capacity to build a room temperature cooled final amplifier, providing few Joules energy of the seed laser pulses with in a 100 s TW/100 s Hz CPA laser systems.

Numerical modeling of scaling larger peak power amplifier modules with double channel cooling and double crystals (three cooling channels) design were also conducted in Ref. [43].

Double channel cooled disks with diameters ranging from 6 to 20 cm and corresponding thicknesses of 0.6 to 2 cm, were investigated (Figure 19a). The inlet flow velocity was 4 m/s in all cases to ensure high levels of heat extraction from the disk gain modules. Every gain module would remain under 45°C up to 40 Hz of operation relative to the coolant temperature. When the repetition rate is growing up further, one can see significant rise in the temperature increase (TI), nevertheless, the temperature profile would remain smooth and flat in the region of the laser amplification. Peak power could reach 8.5 PW and an average power 17 kW with compressor efficiency of 60% and pulse duration of 20 fs [Figure 19a (inset)].

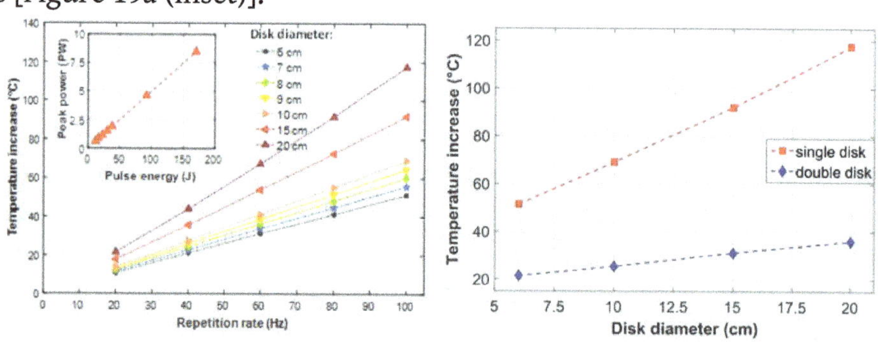

a b

Figure 19. (a) Temperature dependence on repetition rates for various single disk sizes. Pump energy 40, 57, 77, 101, 127, 308, 567 J, for the diameters starting from 6 cm respectively. Peak power of compressed pulses listed in inset (compressor efficiency-60%, repetition rate-100 Hz and pulse duration-20 fs); (b) temperature in the single and double disk modules (Ti:Sa crystals of 6, 10, 15 and 20 cm diameters, 100 Hz for all cases) cooled by three channels using 4 m/s flow velocity at the inlet boundary of the channels.

The maximal repetition rate when amplifiers could operate safely and without serious beam degradation can be estimated based on the obtained TIs. Extremely efficient heat extraction can be obtained by increasing the diameter, while maintaining the aspect ratio of the gain disks with two coolant channels and thus flat temperature profiles with high repetition rate operation.

Further increase of the average power could be achieved by splitting the gain disk to multiple plates with reduced thickness and increasing the number of coolant channels. Four gain modules with double disk sizes ranging from 6 to 20 cm was investigated, with three coolant channel arrangement (Figure 19b). These simulations demonstrate 2 kW output average power with a TI in the disks of 21.5°C, and 28 kW average power at TI of 36°C using multiple disks and cooling surfaces with proper coolant flow conditions.

In summary, this analysis concludes that replacing ordinary thickness Ti:Sa crystals with booster/final amplifiers in ultra-high pack power laser systems using EDP-TD Ti:Sa amplifiers leads to a large improvement in repetition rates at an average power. These systems have the potential to achieve unprecedented parameters, such as 100 terawatts per second per hertz and up to 1 kilowatt of average power output. This can be accomplished utilizing cooling systems that operate at ambient temperature and existing pump lasers that can supply a few joules of energy in the green spectrum at the same repetition rate. Simultaneously, there are no restrictions on increasing these values to a few petawatts of peak power and a repetition rate of 100 Hz once the necessary pump lasers have been developed.

5. CONCISE SUMMARY

This chapter introduced many concepts for enhancing the ultra-high peak power CPA laser systems through innovation. By harnessing these concepts, it becomes possible to greatly enhance the amount of energy produced (reaching levels of kilojoules), decrease the duration of each pulse (down to a few femtoseconds), and thus boost the peak power output to the order of hundreds of petawatts. Simultaneously, the potential for average power increase in these systems up to 10 s kW was also showcased.

The EDP method for Ti:Sa final amplifiers has been identified as the most straightforward approach to achieve a significantly high output energy [25, 27, 28]. When the EDP amplifier is operating under optimal conditions, it has the ability to greatly enhance the extracted energy and minimize the losses associated with TASE and TPG. By utilizing the current wide opening of Titanium:Sapphire crystals and employing liquid absorbers that have the same refractive index, it becomes feasible to achieve energy extraction at a level below one kilojoule. By achieving a compressor transmission efficiency of 70% and a pulse duration of 15 fs, it is possible to attain a power level of around 30 PW. The efficacy of EDPCPA technology has been demonstrated through its implementation in numerous world-class laboratories, resulting in an output energy of around 200 J and a record-breaking peak power of 5 PW. The existing Ti:Sa crystals, with a diameter of 20-30 cm, can be used to achieve an output energy of around 500-800 J in the next steps.

Subtitle 3 discusses two recently established methods of pulse shortening. The study demonstrated that it is possible to achieve a significantly wider range of wavelengths in Ti:Sa laser amplifiers by utilizing both π- and σ-axis and manipulating the spectral polarization of the amplified pulses [34]. A proof-of-principle experiment produced an amplification bandwidth of over 85 nm with a gain of 200. These experiments have demonstrated that actively pre-shaping the pulse spectrum with pre-amplified amplification before saturated amplification in conventional CPA amplifiers can effectively counteract the spectral red-shifting and gain narrowing that occur during amplification in Ti:Sa CPA systems. The computer modeling shown that a polarization-encoded chirped pulse amplification approach may be expanded to higher energy levels and generate multi-Joule pulses with a bandwidth of around 200 nm. This advancement makes it possible to create few-cycle petawatt Ti:Sa systems.

The spectral widening technique utilizing self-phase modulation (SPM) in the material's bulk, along with multi-stage compression and subsequent recompression of the chirped pulse, enables the generation of ultra-short pulse durations below 10 fs, without any loss of energy [37]. The suggestion was made to further expand this notion by using SPM in thin film below 1 mm and achieving significantly higher intensity, reaching up to tens of TW/cm^2 [38]. Numerical simulations were conducted on a two-stage thin film compressor using thin film elements with thicknesses of 0.5 mm and 0.1 mm. The projected peak intensities following the first and second stages, using the temporal recompression approach, are 16.6 and 43 TW/cm^2, with pulse lengths of 6.4 and 2.1 fs, respectively. By reducing the duration of the pulse without any loss of energy, it becomes possible to significantly boost the peak power output by an extra magnitude, resulting in the ability to achieve power levels in the range of several hundred petawatts from a single channel of CPA laser systems.

The integration of EDP technology with TD Ti:Sa crystals for power amplifiers [39, 41, 43] enables the enhancement of both the ultra-high peak power and the average power. A proof-of-principle experiment successfully demonstrated high-energy broadband amplification using a room temperature water-cooled EDP-TD head. This was achieved at a repetition rate of 10 Hz, without the need for a typical cryogenically cooled multipass method. Hence, the EDP-TD combination can surpass the constraints related to thermal effects and transverse amplified spontaneous emission, hence allowing Ti:Sa laser systems to achieve a peak power of petawatts and repetition rates in the hundreds of hertz, or average power in the kilowatts range.

AT A GLANCE

Ultra-high peak power laser systems have potential applications in emerging and very promising fields, such as accelerating charged particles and confining the fusion nuclear reaction using inertial confinement. The first one can be utilized as a highly efficient secondary source of Y and X-ray beams, which have several applications in both industry and medicine. By increasing the repetition rate, its effectiveness can be enhanced. On the other hand, the last one has the potential to function as an inexhaustible energy source once it is converted into power plants. Emerging technologies have the capability to greatly enhance the maximum power output by extracting bigger amounts of energy during the pumping process, known as energy extraction during pumping (EDP). This technology achieved a record of extracting 200 joules of energy and producing an output power of 5 petawatts. This chapter introduces the Polarization Encoded Chirped Pulse Amplification (PE-CPA) approach, along with two steps of compression, which result in a reduction in pulse duration. In addition, this study will showcase the potential of combining the EDP technique and the Thin Disk (EDP-TD) in Ti:Sa amplifiers to achieve greater repetition rates in PW-class laser systems. Furthermore, the results of proof-of-principle tests will be shown.

REFERENCES

1. McClung F et al. Giant optical pulsations from ruby. Journal of Applied Physics. 1962;33(3):828
2. Mocker W et al. Mode competition and self-locking effects in a Q-switched ruby laser. Applied Physics Letters. 1965;7:270
3. Tajima T et al. Laser electron accelerator. Physical Review Letters. 1979;43:267
4. Nicollas et al. Lasers and controlled thermonuclear fusion. Nature. 1972;239:1-29
5. Available from: https://lasers.llnl.gov/about/nif [Internet]
6. Available from:http://www-lmj.cea.fr/index.html [Internet]
7. Damm T et al. Compression of picosecond pulses from a solid-state laser using self-phase modulation in graded-index fibers Optics Letters. 1985;10(4):176
8. Strickland D et al. Compression of amplified chirped optical pulses. Optics Communication. 1985;56(3):219
9. Giordmaine J et al. Compression of optical pulses. Journal of Quantum Electronics. 1968;QE-4:252
10. Duguay M et al. Compression of pulses from a mode-locked He-Ne laser. Applied Physics Letters. 1969;14:14
11. Treacy E. Optical pulse compression with diffraction gratings. IEEE Journal of Quantum Electronics. 1969;5(9):454
12. Martinez O et al. Negative group-velocity dispersion using refraction. Journal of the Optical Society of America A. 1984;1(10):1003
13. Moulton P. Spectroscopic and laser characteristics of Ti:Al2O3. Journal of the Optical Society of America B. 1986;3:125

14. Taiha Joo, Yiwei Jia, Graham R. Fleming, Ti:sapphire regenerative amplifier for ultrashort high-power multikilohertz pulses without an external stretcher. Optics Letters. 1995;20:389

15. Ian N. Ross et al. Generation of terawatt pulses by use of optical parametric chirped pulse amplification. Applied Optics. 2000;39(15):2422

16. Ple F et al. Design and demonstration of a high-energy booster amplifier for a high-repetition rate petawatt class laser system. Optics Letters. 2007;32(3):238

17. Gerstner. Extreme light. Nature. 2007;446:16

18. Available from: http://www.eli-laser.eu [Internet]

19. Yamakava K et al. Generation of 16-fs, 10-TW pulses at a 10-Hz repetition rate with efficient Ti:sapphire amplifiers. Optics Letters. 1998;23(7):525

20. IshII N et al. Multimillijoule chirped parametric amplification of few-cycle pulses. Optics Letters. 2005;30(5):567

21. Mourou G et al. Optics in the relativistic regime. Reviews of Modern Physics. 2006;78(2):309

22. Speiser J. Scaling of thin-disk lasers—influence of amplified spontaneous emission. Journal of the Optical Society of America B: Optical Physics. 2009;26:26

23. Peterson F et al. Suppression of parasitic lasing in large-aperture Ti:sapphire laser amplifiers. Optics Letters. 1999;24(14):963

24. Bonlie J et al. Production of $>10^{21}$ W/cm^2 from a large-aperture. Ti:sapphire laser system. Applied Physics B. 2000;70:155

25. Chvykov V et al. Transverse amplified spontaneous emission: the limiting factor for output energy of ultra-high power lasers. Optics Communication. 2014;312:216

26. Koechner W. Solid State Laser Engineering. Berlin: Springer; 1996. p. 747

27. Chvykov V et al. Suppression of parasitic lasing in multi-pass Ti-sapphire amplifiers, CWA34. In: Technical Digest CLEO. 2003

28. Chvykov V et al. Large aperture multi-pass amplifiers for high peak power lasers. Optics Communication. 2012;285(8):2134

29. Joyce D et al. Ti:sapphire laser crystals reach 200 mm diameter. Laser Focus World. 2010;46(3):37

30. Chu Y et al. High-energy large-aperture Ti:sapphire amplifier for 5 PW laser pulses. Optics Letters. 2015;40(21):5011

31. Yanovsky V et al. Ultra-high intensity-300-TW laser at 0.1 Hz repetition rate. Optics Express. 2008;16:2109

32. Sung JH, Lee SK, Lee HW, Yoo JY, Nam CH. Development of 0.1 Hz 4.0 PW laser at CoReLS. In: Conference on Lasers and Electro-Optics, OSA Technical Digest (2016) (Optical Society of America, 2016), paper SM1M.3

33. Chvykov V et al. Final EDP Ti:Sapphire amplifiers for ELI project. Proceedings of SPIE. 2015;9515 OJ. DOI: 10.1117/12.2182319

34. Kalashnikov M et al. Polarization-encoded chirped pulse amplification in Ti:sapphire: A way toward few-cycle petawatt lasers. Optics Letters. 2016;41(1):25

35. Claude Rolland, Corkum PB. Compression of high-power optical pulses. Journal of the Optical Society of America B. 1988;5:641

36. Mevel E et al. Multielectron ionization processes in ultrastrong laser fields. Journal of the Optical Society of America B. 2003;20:105

37. Chvykov V et al. Compression of ultra-high power laser pulses. CLEO-Proceeding. 2010: JThG4

38. Mourou G et al. Single cycle thin film compressor opening the door to zeptosecond-exawatt physics. The European Physical Journal Special Topics. 2014;223(6):1181

39. Chvykov V et al. Design of a thin disk amplifier with extraction during pumping for high peak and average power Ti:Sa systems (EDP-TD). Optics Express. 2016;24(4):3721

40. Negel J, et al. 1.1 kW average output power from a thin-disk multipass amplifier for ultrashort laser pulses. Optics Letters. 2013;38:5442

41. Chvykov V et al. High peak and average power Ti:sapphire thin disk amplifier with extraction during pumping. Optics Letters. 2016;41(13):3017

42. Bahk S et al. Generation and characterization of the highest laser intensities 1022 W/cm2. Optics Letters. 2004;29:2837

43. Nadymihaly R et al. Liquid-cooled Ti:Sapphire thin disk amplifiers for high average power 100-TW systems. Optics Express. 2017;25(6):6664

CHAPTER-5

HIGH-POWER THIN-DISK OSCILLATORS EMPLOYING KERR-LENS MODE-LOCKING

1. INTRODUCTION: AN OVERVIEW

Femtosecond lasers can be categorized based on many output factors, including pulse duration, repetition rate, average power, peak power, carrier envelope phase stability (CEP), intensity noise, and long-term stability. Depending on the specific application, just a select few factors from that list may be of importance. The selection of a specific laser or oscillator technology is typically influenced by the intended uses, and ultimately, it is the users and the market that determine which technology is advantageous for each application. Prior to commercially pursuing a certain technology, researchers dedicate significant efforts to demonstrate its potential for various applications. The current situation pertains to the thin-disk (TD) oscillators. Currently, commercially available technologies such as femtosecond slab amplifiers [1] and fiber amplifiers [2] are functioning in the MHz repetition rate range and are already being used to enable practical applications. Thin-disk oscillators offer a more compact and simplified alternative to amplifiers, and they possess certain features that can be preferable than amplifiers.

The authors' initial motivation was to conduct research in two distinct spectral ranges: extreme ultraviolet (XUV) and medium infrared radiation (MIR). These laser sources would ideally offer access to both wide-spectrum radiation in the XUV (extreme ultraviolet) range of 13-200 nm and wide-spectrum radiation in the infrared range of 3-30 µm. Thin-disk oscillators possess inherent simplicity and compactness, making them suitable as table-top substitutes for synchrotron beamlines [11, 12]. In addition, this broadband radiation has the potential to be converted into frequency combs [13, 14] with the use of additional stabilizing measures [5], and can potentially serve as a compact source for generating attosecond pulses [15, 16] on a table-top scale. Therefore, the advancement of direct frequency comb or dual frequency comb spectroscopy in the XUV and MIR regions would be substantial compared to the existing level of technological development. Furthermore, these sources are portable, which would facilitate additional experiments and partnerships within the research community.

The rapid advancement in the creation of Yb-doped lasers was primarily facilitated by the accessibility of cost-effective pump sources, specifically InGaAs diodes [17]. The price of these pump diodes has consistently declined in recent years as a result of growing demand in industrial applications. Given the present cost of pump laser diodes, the optical to optical efficiencies of femtosecond oscillators, even at moderate levels, are not a significant worry, particularly for research purposes. The introduction of Kerr-lens mode-locking [18] in conjunction with Ti:Sapphire (Ti:SA) [19] as the amplifying material had a significant influence on applications, both in research and beyond. Upon closer examination of the advancements in Ti:Sa and Yb:YAG thin-disk oscillators, it can be inferred that the Yb-based TD oscillators bear resemblance to the technological development of the Ti:Sa oscillators. The primary benefit of Yb-based TD technology is in its capacity to scale both peak and average power, utilizing cost-effective pump diode sources. Nevertheless, the Ti:Sa gain medium possesses an exceptional broadband characteristic and emits light inside the visible spectrum, which is typically not easily achievable with Yb-based gain media.

Therefore, Yb-based TD technology is not a complete replacement for Ti:Sa technology, particularly in applications that do not require high average powers. Furthermore, once direct diode pumping of Ti:Sa oscillators becomes the standard pumping method, we can anticipate a future competition between Yb-based and solid-state Ti:Sa oscillators, particularly in the low average power range. The thin-disk concept is extremely important for applications that use XUV and MIR radiation and require scalability. For example, when we consider basic intra-pulse difference frequency production, the conversion efficiencies from a 1 μm driving source into the XUV are often less than 10-5 [20], and less than 10-3 into the MIR [21]. Therefore, in order to achieve the required power levels for spectroscopy applications, it is necessary to utilize driving lasers with very high average powers, often ranging from 100 to 1000 W. Femtosecond enhancement cavities [22, 23, 24] are a form of technology that is very suitable for studies involving XUV generation. Currently, direct XUV frequency comb spectroscopy has only been proven using this very sophisticated technology [25], and the highest average powers have been obtained.

We analyze femtosecond oscillators that produce average powers above 100 W, with pulse durations of around 200 fs and repetition rates ranging from a few tens of MHz. For more details, please refer to Table 1, which provides an overview of several oscillators. These parameters are acquired directly from oscillators without the need for any form of extra amplification. Thin-disk oscillators are a distinct type of lasers that have the exceptional ability to combine both high peak and high average powers. The primary characteristics of these devices include their lack of amplification, minimal noise, and very small size. Subsequently, we exclusively concentrate on this specific technology, disregarding any amplifiers or augmentation cavities.

Table 1. Summary table of the KLM thin-disk oscillators developed at MPQ, LMU and UFI GmbH from 2012 till 2017.

P_{avg}, W	E_p, μJ	f_{rep}, MHz	τ_p, fs	P, MW	P_{avg}, W (compressed)	τ_p, fs (compressed)	Application	Status
40	3	13	300	9			Seed oscillator	In use [3]
90	0.9	100	250	3.5	>50	20	MIR generation	In use [4]
42	1.1	38	250	4.2	6 (10)	7.7 (10)	MIR generation	In use [5, 6]
270	14	19	330	37.8			Development itself	Not in use [7]
155	10	15.5	140	63	130	30	XUV generation, Raman spectroscopy	In use [8]
10 (3.5)	0.7 (0.4)	100–200	70 (47)	0.6			Development itself	In use [9]
100	4.1	24	190	19.3	65	30	Commercial system	In use [10]

Yb:YAG is the preferred gain medium for all oscillators. The majority of the oscillators are effectively functioning in the laboratories using the parameter sets that were initially released.

In 2000, Prof. Keller's group successfully demonstrated the first mode-locked Yb:YAG thin-disk oscillator [27]. The paper effectively combined two existing technologies: the thin-disk and Semiconductor Saturable Absorber Mirrors (SESAM). Over the following decade, the same group made more progress in this technology and achieved numerous milestones in terms of maximum power, pulse energy [28], and average power [29] derived directly from femtosecond oscillators. In addition, several research groups have explored these boundaries [30, 31] and examined other gain materials and dispersion regimes [32]. However, the combination of the Kerr-lens mode-locking (KLM) technique with the TD technology was not shown until 2012 [33]. While the concept of combining these two technologies was patented in 1999 [34], the practical efforts to achieve it were deemed fruitless as stated in [35]. Our group successfully demonstrated the merger of KLM and TD technologies in 2012 [33]. This initial promising trial served as a catalyst for us to continue exploring this path.

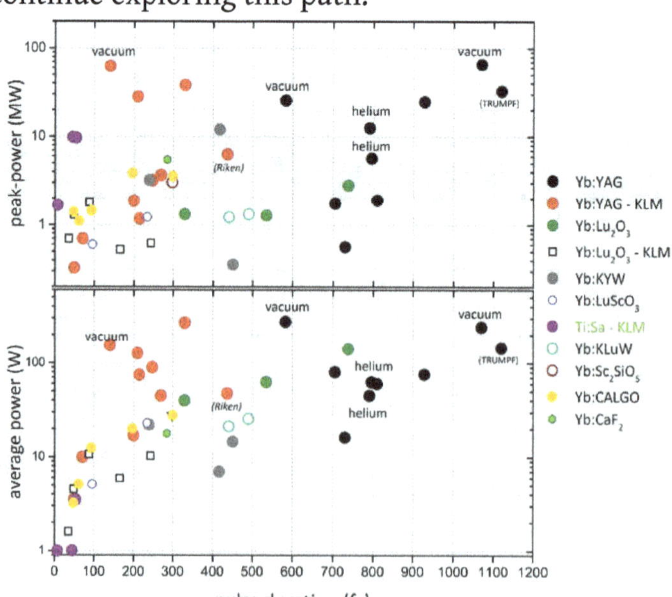

Figure 1. Summary of diverse femtosecond thin-disk oscillators adopted from [26]. More details including the references can be found in [26].

In the past 5 years, several TD KLM oscillators have been constructed, with additional organizations also participating in this endeavor [36, 37, 38]. The oscillators created by our research team are outlined in Table 1 and are visually represented as red dots in Figure 1. Furthermore, Figure 2b illustrates the fast development of TD KLM oscillators.

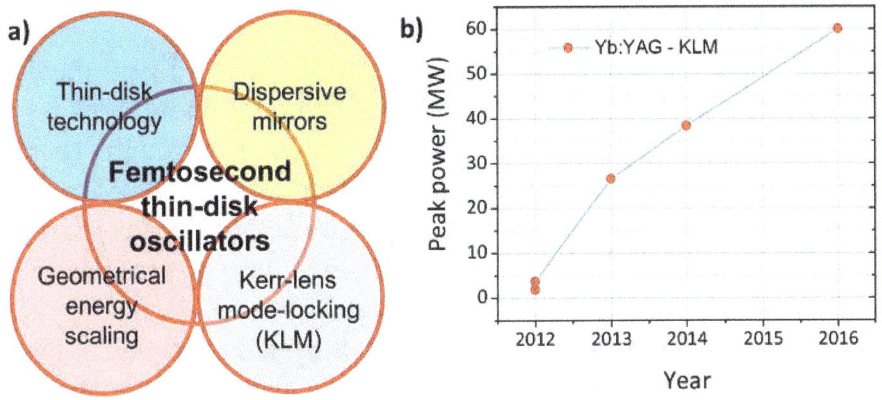

Figure 2. (a) Four key elements: thin-disk technology, dispersive mirrors, Kerr-lens mode-locking and geometrical energy scaling concept form femtosecond thin-disk oscillator technology described and (b) graphical representation of the rapid thin-disk KLM oscillator development in our group.

2. KERR-LENS MODE-LOCKING PRINCIPLE

The refractive index n of a material depends on the incident electric field intensity. A Gaussian intensity distribution causes an increase of the refractive index in the central part of the beam relative to its outer regions therefore forming a nonlinear lens. The higher the light intensity, the stronger the action of such a lens. The lens becomes stronger for smaller beam radii ω and media with higher nonlinear refractive index n 2. Self-focusing occurs for a high-intensity, pulsed laser-beam (red, Figure 3) and reduces losses due to the hard aperture blocking the continuous wave (CW) beam of lower intensity. In a resonator-cavity, this mechanism initiates mode-locking and acts as an artificial saturable absorber. Catastrophic run away damage can happen when a critical power is reached and the length of the medium exceeds the self-focusing length. The first oscillator working on the KLM principle was discovered by Spence et al. [18] and referred to as self-mode-locking or magic mode-locking. Piche [39] explained the mode-locking mechanism on the basis of self-focusing and only a few authors recognized the potential of the self-focusing effect for mode-locking before the invention of KLM [40, 41]. Since that time, KLM established itself as the method of choice for ultrashort-pulse generation and numerous studies were done on resonator design, theoretical numerical and analytical description of KLM and experiments on ultrashort pulse generation.

Mostly, experiments were performed with the Ti:Sa gain medium, which has several outstanding features: an extremely broad gain bandwidth, short upper-state lifetime as well as high thermal conductivity [19]. Understanding that shortest possible pulses can only be obtained when nonlinearities and dispersion are balanced to form so-called soliton pulses [42, 43] preceded the invention of KLM. However, this technique constituted the decisive building-block to enable robust, usable solid-state femtosecond oscillators. With Kerr-lens mode-locked solitonic Ti:Sa oscillators up to several 100 mW average power and up to MW-level peak-power were realized with pulse durations approaching few optical cycles, all in a compact, reliable setup that was superior to the old dye-based technology. This ensured its worldwide adoption in many optical laboratories and nearly immediate commercialization.

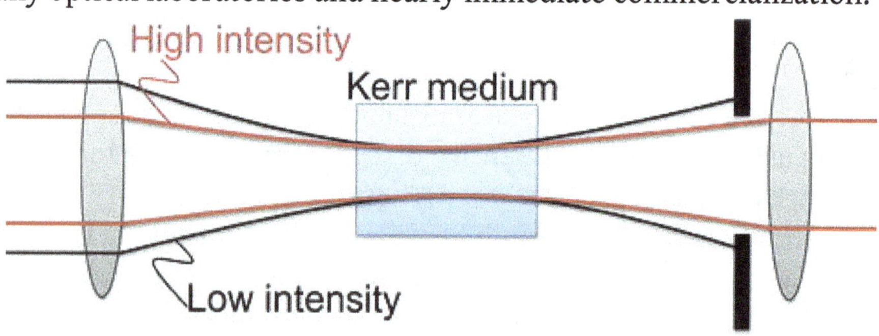

Figure 3. Basic principle of KLM. Self-focusing occurs for a high-intensity beam (red) and reduces the losses due to a hard aperture (two black knifes) blocking the low-intensity (CW) beam. This mechanism initiates mode-locking and acts as an artificial saturable absorber.

3. POWER SCALING IN KLM OSCILLATORS

For power scaling of soliton mode-locked oscillators both the nonlinear and the thermal instabilities need to be considered. A thin-disk gain medium is in both cases an ideal solution. The predominantly one-dimensional heat flow within the thin gain medium minimizes thermal lensing and the low peak intensities in the large but thin disk minimize the nonlinear phase shift. Unlike the usual bulk KLM oscillators [44], high power thin-disk KLM oscillators require that the role of the gain and Kerr-medium are separated to distinct intra-cavity elements. This way it is possible to keep the mode size on the thin-disk large and to tune the nonlinear phase shift nearly independently by means of an additional Kerr-medium (see Figure 4a).

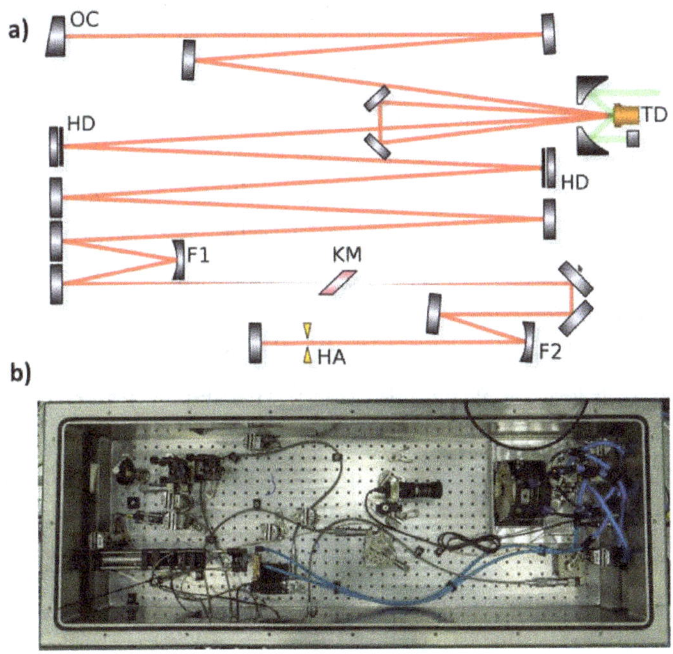

Figure 4. (a) Oscillator sketch. TD: thin-disk; OC: output-coupler; HD: highly dispersive mirrors; $F_{1,2}$: focusing mirrors; KM: Kerr-medium; HA: hard aperture and (b) a photo of a 100 W-level thin-disk oscillator in a compact monolithic aluminum housing. Photo credit: Kilian Fritsch.

3.1. Average power scaling

One of the crucial advantages of KLM compared to semiconductor saturable absorbers is nearly negligible linear and nonlinear absorption inside of a dielectric Kerr-medium. SESAMs normally exhibit non-saturable losses and multiphoton absorption [45]. Such low absorption losses are essential for average power and peak power scaling. Thus, in 2014, quickly after the first demonstration of thin-disk KLM [33] up to 270 W average power from a KLM thin-disk laser were demonstrated [7] (also see Figure 2b). The utilized thin-disk technology remains the key component for high average power operation for both Kerr-lens and SESAM mode-locking techniques. While the first demonstrator of a KLM thin-disk oscillator [33] relied on a copper mounted thin-disk the record performances approaching 300 W output were achieved with a thin-disk contacted to a diamond heat sink [7]. The superior heat-conductivity of diamond in combination with a suitable contacting technique [46] allow both larger pump intensities as well as thinner disks to be used which helps to reduce thermal lensing and aspherical aberrations [47].

Another important aspect is thermal lenses in the oscillator elements. KLM oscillators require an increased sensitivity to the Kerr-lens to initiate mode-locking and retain a strong self-amplitude modulation (SAM) effect. This higher sensitivity is achieved by adjusting the resonator close to the edge of stability [48, 49] where, however, not only the sensitivity to the Kerr-lens but also to any thermal lens inside the resonator is increased. In particular, the Kerr-medium material has a very pronounced influence on the oscillator behavior. For intra-cavity average power levels larger than 500 W, the use of fused silica plates is difficult at best, exhibiting strong thermal lensing. This is observable, e.g., by shifting the Kerr-medium under CW irradiation along the beam-axis and monitoring the mode-size change. Suitable alternatives are found in sapphire (46 W m^{-1} K^{-1} [50]), crystalline quartz (10 W m^{-1} K^{-1} [50]) or YAG (13 W m^{-1} K^{-1} [50]) materials, all having a significantly higher thermal conductivity than fused silica (1.4 W m^{-1} K^{-1}). With a sapphire Kerr-plate more than 1.2 kW, stable average power could be demonstrated inside the KLM oscillator.

Furthermore, highly dispersive mirrors were found to exhibit additional thermal lensing and to prevent average power scaling [47, 51, 52]. However, after a few attempts in manufacturing of those mirrors, it was possible to identify a multilayer design showing no thermal effects and still providing significant dispersion levels around −3000 fs^2 within 10 nm spectral bandwidth [53]. Another thermal lens suppression method implies the use of substrates with higher thermal conductivity and (or) lower thermal expansion coefficients [54, 55].

3.2. Peak power scaling: general aspects

The pulses traveling inside the KLM oscillator cannot take on arbitrary pulse energy E and width T but have to obey the well known soliton propagation law. It is assumed that soliton mode-locking is stable at a certain peak power level P_0 [56] then

$$T = \sqrt{\frac{|\beta_2|}{\gamma P_0}} \tag{1}$$

with the group delay dispersion (GDD) β_2 and the nonlinear self-phase modulation (SPM) coefficient γ . The pulse can retain its shape when the frequency chirp due to SPM is canceled by (anomalous) dispersion of opposite sign. Pulses with a peak-power smaller or larger than P_0 do not meet this condition. If the peak-power is too small, the pulses are dispersed and if it is too high, several instabilities can occur. Commonly observed instabilities are the splitting into two or more pulses per roundtrip period or the appearance of additional CW-background radiation. This equation describes well the case not only when pulses travel in lossless media such as fibers or a passive resonator but also catches the dilemma of power-scaling in mode-locked oscillators.

The pulse energy of a soliton can be increased, e.g., by introducing larger values of the GDD. New coating methods allow introducing GDD on the order of 100,000 fs^2 with highly dispersive mirrors [7, 30, 57]. However, the larger chirp from the strong dispersion is automatically compensated by a pulse-duration growth that leaves the peak power constant. This way highest pulse energies were generated and extracted from SESAM mode-locked oscillators at the expense of longer pulses [28, 30].

The second important variable in Eq. (1) is the reduction of the SPM parameter which depends on the effective mode-area A_{eff} inside the oscillator elements of thickness d and nonlinear refractive index n_2 . Interestingly, a contribution of air to the SPM parameter cannot be neglected anymore at a certain intra-cavity peak power and cavity length.

$$\gamma \propto \frac{d\, n_2}{A_{eff}} \tag{2}$$

A decrease of γ is therefore directly coupled to the resonator geometry, regardless of whether SESAM or a Kerr-lens is used for mode-locking. The thin-disk gain medium favors large beam areas and short crystal lengths for average power scaling which implies also the minimization of the SPM-coefficient γ . Most state of the art, mode-locked thin-disk lasers are operated in a vacuum or reduced pressure environment to further decrease γ [8, 28, 36]. SESAM mode-locked oscillators are especially sensitive to SPM since their modulation-depth and -speed is intrinsically limited for high-power operation. Current limitations to power scaling in such oscillators are assumed to originate partly from the residual nonlinearity in the mirror coatings [45]. KLM oscillators on the other hand appear to have a much higher tolerance to nonlinear phase shifts being attributed to the achievable large modulation depth as well as the near instantaneous response time of the Kerr-nonlinearity. Thus, fairly large peak-powers can be generated even under normal atmosphere [7].

3.3. Geometrical approach to peak power scaling

Increasing the peak-power in KLM oscillators appears straightforward using a geometrical approach. A fair assumption is that the main contribution to the pulse nonlinear phase shift originates from the interaction in the Kerr-medium. Therefore, in an oscillator where all parameters such as dispersion, self-amplitude modulation, losses, pulse duration, etc. are fixed and only the mode area in the Kerr-medium is increased, the peak power should increase linearly with A eff . This is not in contradiction with the soliton condition as stated in Eq. (1). The Kerr-medium is located in the beam waist of a focusing arrangement as seen in Figure 4. The mode radius w kerr in the Kerr-medium is linearly dependent on the curvature radius R 1 , 2 of the focusing arrangement mirrors F1 and F2. Therefore, varying R 1 , 2 by a factor N , one would expect to increase the peak power favorably by the factor N 2 . A necessary condition for this is that SAM also stays constant. This condition, however, is not necessarily fulfilled. Unfortunately, the decisive SAM parameters of a certain KLM resonator such as modulation-depth or saturation-power are not easily measurable and even simulations appear challenging due to the coupling between temporal and spatial extent of the pulses. The dependence of SAM on the resonator geometry therefore gives an experimentally observed deviation from the initially expected, quadratic scaling of the peak power.

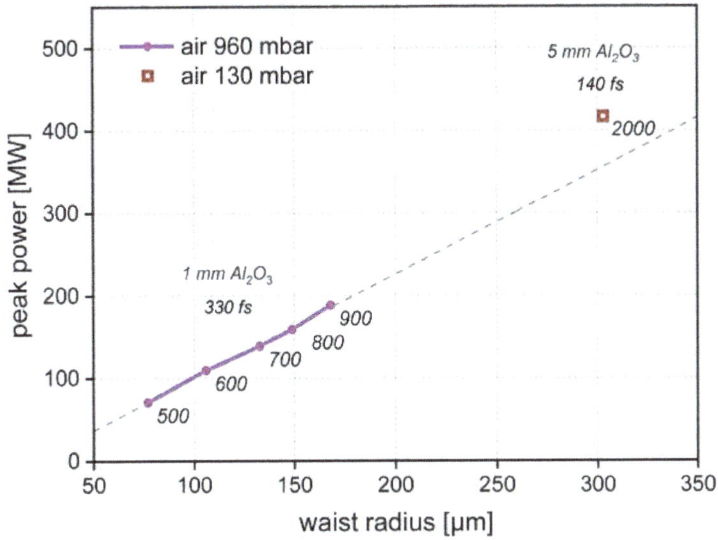

Figure 5. (a) Summary of the data from the geometrical KLM power-scaling approach. The waist radius in the Kerr-medium, growing with $R_{1,2}$, is approximated from ABCD-matrix calculations and (b) experimental example of intra cavity peak power and pulse duration when the total oscillator second order anomalous dispersion (GDD) is increased.

The geometrical power scaling concept was applied experimentally as published in earlier work of the authors [7, 8]. The KLM oscillators were operated in air both under normal and reduced pressure environments. For the scaling experiments in work [7], all parameters except the pump power were kept constant. Four passes of the laser mode through the Yb:YAG thin-disk (TRUMPF Laser GmbH) per roundtrip allowed to couple out large fractions of the intra-cavity power (21% transmission of the output coupler). The mode size inside the 1 mm thick sapphire Kerr-plate was varied by successively exchanging the mirrors $F_{1,2}$ to have R 1 , 2 ranging from 300 to 900 mm. The resulting peak power increased proportionally to R 1 , 2 . These results are summarized in Figure 5.

As described previously, the soliton peak-power is fairly invariant under a change of the intra-cavity GDD. However, the changes in pulse energy and duration can be substantial and give flexibility in the design parameters. This is demonstrated in Figure 5b, where the GDD in the thin-disk KLM oscillator was varied from -18,000 to -48,000 fs^2 by exchanging dispersive mirrors. In particular example, the intra-cavity pulse energy increased from 32 to 57 μJ while the pulse duration also increased from 210 to 330 fs showing little effect on the peak power.

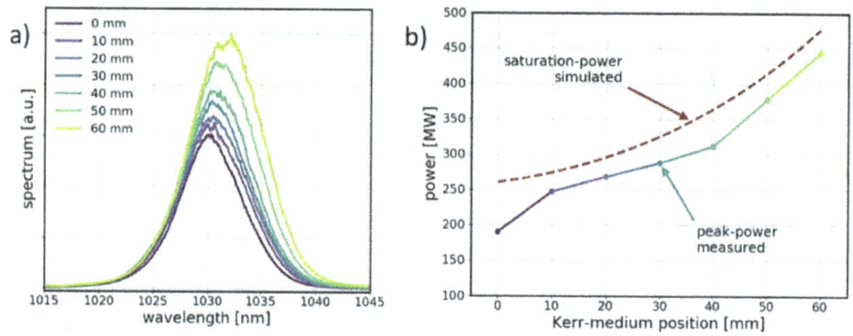

Figure 6. (a) Spectral broadening and spectral shift for different Kerr-medium positions with respect to the focus and (b) intra-cavity peak power for different Kerr-medium positions. The saturation-power of the SAM is also estimated from ABCD-calculations and plotted.

Soliton mode locking with strong self-phase modulation allows the direct generation of pulses approaching the spectral gain bandwidth of the laser-medium. SPM plays a key role in replenishing those spectral components of the pulse which do not see sufficient net amplification from the gain emission spectrum. A reasonably small value of the intra-cavity dispersion as well as a strong self-amplitude modulation is necessary to reach this regime. In a KLM oscillator combining a fairly low roundtrip GDD of -12,000 fs^2 as well as a sapphire plate of 5 mm thickness, 140 fs pulses could be generated with a high optical-to-optical efficiency of 29%. The oscillator operated with a focusing section incorporating mirrors R 1 , 2 with 2 m curvature. This allowed an intra-cavity peak-power level of more than 400 MW with more than 60 MW output peak power. These results are in a good agreement with the linear power-scaling curve in Figure 5a.

The SAM is quite sensitive to the position of the Kerr-medium within the focus of $F_{1,2}$. Simulations show that mostly the saturation power is affected which intuitively follows from the change of intensity in the medium. Data taken for a Kerr-medium being translated through the focus are displayed in Figure 6. It is evident that both the spectral bandwidth and the peak-power increase while the Kerr-medium position is shifted out of the focus. A factor of nearly 2 could thus be gained in intra-cavity peak-power, from near 200 MW to more than 400 MW. Beyond this point mode-locking could not be observed.

3.4. Influence of atmosphere

The relatively short (~ 5 m) KLM oscillators with R 1 , 2 up to 1 m are fairly insensitive to pressure changes of the encompassing air and most of the nonlinear phase-shift is acquired in the Kerr-medium. The increase of Rayleigh distance for weaker focusing (longer R 1 , 2) leads to the increase of nonlinear phase acquired in gas. Thus, the contribution of air relative to that from the bulk Kerr-medium becomes significant. This is potentially harmful as (i) the pulse-stability becomes more dependent on air-fluctuations due to the coupling to SAM and SPM (ii) the oscillator SAM cannot be optimized independently anymore by positioning of the Kerr-medium as the atmosphere begins contributing to the mode-shaping.

This effect can only be counteracted by evacuating or decreasing the pressure of the atmosphere in the oscillator. A rough estimation of the fractional contribution of the bulk Kerr-medium to total nonlinear phase-shift, depending on waist-size of the cavity-mode and different pressure levels is depicted in Figure 7.

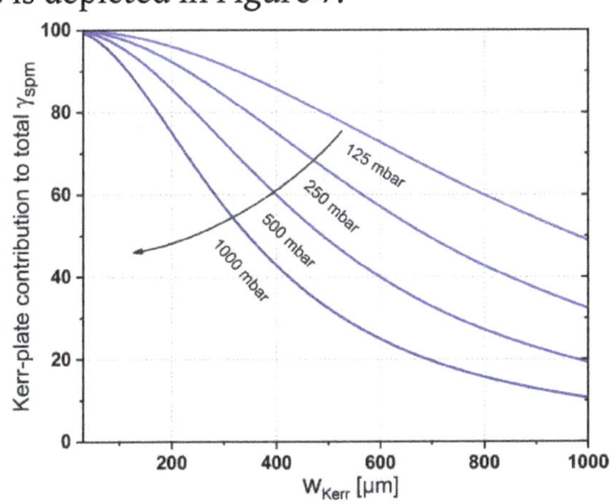

Figure 7. Estimated contribution of the Kerr-plate to the total nonlinearity γ, originating from both the solid Kerr-plate and the air inside the oscillator. It is plotted for different air-pressures.

For these reasons, it was necessary to reduce the atmospheric pressure in the oscillator from [8] to about 150 mbar for stable operation whereas the effect of evacuation proved negligible in the short- R 1 , 2 oscillators [7]. The total nonlinear phase-shift acquired in the reduced-pressure oscillator with R 1 , 2 being 2 m was approximately 0.6 rad. In that case, the contribution of air to the total nonlinear phase-shift can be estimated to stay favorably below 10%.

Although the geometrical peak power scaling concept described here exhibits a seemingly linear dependence of the intra-cavity peak-power on the mode size in the Kerr-medium, it is very interesting to further verify this dependence for even larger mode sizes and higher intra-cavity peak powers.

3.5. Intensity noise and CEP stabilization

Since power-fluctuations can be enhanced in nonlinear processes like SPM, high-harmonic generation, difference frequency generation or optical parametric amplification it is necessary that the pulses generated from the oscillator are as noiseless as possible. It is not obvious that high-power oscillators, especially those with long cavities, can be as silent as usual low power bulk KLM oscillators. Concerns were also raised by some that high-power KLM oscillators would also suffer from instabilities since the initiation of mode-locking requires a setup close to the cavity stability limit. State of the art KLM thin-disk oscillators deliver more than 100 W average power and, thus, are by two orders of magnitude more powerful than usual KLM bulk oscillators. For instance, the oscillator described in the previous section [8] delivers 150 W average and 60 MW peak power.

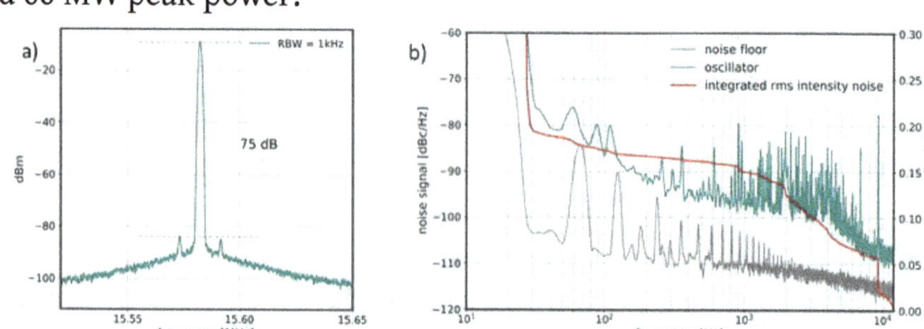

Figure 8. (a) RF signal around the repetition rate of the oscillator with 60 MW output peak power. Small sidebands are visible more than 75 dB below the signal and (b) low-frequency noise performance.

The intensity noise of this oscillator was characterized (see Figure 8). Measurement of the laser output with an RF-signal analyzer reveals that nearly all of the noise is generated in the low-frequency band below 10 kHz. This can be attributed to both technical noise such as water-cooling, pump-source or air-turbulences as well as gain-relaxation dynamics. The latter lies in the lower kHz range due to the approximately 1 ms upper state lifetime of Yb:YAG. An intensity noise of 0.3% rms deviation in the 1 Hz–1 MHz window is typical for KLM thin-disk oscillators. At slightly reduced power levels even better values can be measured on a daily basis. These intensity-noise values are comparable to commercial Ti:Sapphire oscillators and promise good results for CEP stabilization of high peak and average power thin-disk oscillators.

Many advanced scientific applications such as high precision spectroscopy in the XUV-VUV range or attosecond pulse generation require the lasers to be CEP stabilized. Obviously, the demonstration of CEP stabilization of femtosecond thin-disk oscillators is an important step towards enabling these applications with a compact, transportable thin-disk oscillator source. Moreover, a low intensity noise is a critically important prerequisite for the CEP stabilization demonstration. The first carrier envelope offset frequency stabilization of a femtosecond thin-disk oscillator was demonstrated with a KLM thin-disk oscillator providing a remarkably high average power of 45 W and a peak power over 4 MW [58]. A similar demonstration followed with a SESAM mode-locked thin-disk oscillator, providing a moderate output power of only 2.1 W [59]. So far, CEP stabilization of high average and peak power oscillators was demonstrated only for KLM thin-disk oscillators resulting in sub-300 mrad out-of-loop noise and output average powers in the 40–50 W range. This was achieved by implementing two CEP control methods, namely intra-cavity loss control by means of an acousto-optic modulator (AOM) [5] and pump-diode control by means of dual-wavelength pumping [60]. Very recently, a new intra-cavity loss modulation approach comprised of an AOM simultaneously acting as a Kerr-medium was implemented [61]. This resulted in the highest average power CEP stabilized laser delivering over 100 W output power and over 30 MW peak power. This approach seems to be scalable by at least an order of magnitude in terms of peak power making it a method of choice for the next-generation CEP-stabilized KLM thin-disk oscillators.

4. DISCUSSION

4.1. Further power scaling opportunities

Even though the geometrical scaling-procedure enables near 100 MW pulses directly from KLM oscillators, it relies on a substantial intra-cavity peak- and average power increase. This raises the demands on the coating damage thresholds during stressful events such as the transition to mode locking. The elevated intra-cavity average powers of more than 1 kW also require careful selection of the utilized materials to prevent thermal lensing instabilities. Complementary to the intra-cavity power-scaling, it is possible to boost the oscillator output by enlarging the output-coupling ratio. This concept has been realized with a mode-locked thin-disk oscillator in [62] for the first time using an imaging multi-pass configuration. Up to 72% of the circulating power were extracted from the SESAM mode-locked oscillator in [30] resulting in 145 W average power output while the intra-cavity value was only 200 W. The short interaction length in the thin disk cannot replenish such high power-loss during a single roundtrip which needs to be overcome by an increased number of beam-passes through the disk, e.g., with an imaging multi-pass cell (20 passes realized in [30]).

The number of disk passes cannot be made arbitrarily high, however, since any thermal lens in the disk is accumulated, giving rise to a narrowing of the cavity-stability zones with respect to the pump power. While this effect has obviously not hindered comprehensive implementation in an oscillator-cavity working at the center of the stability-zone [30], it is not as obvious that the same can be done for KLM-cavities that are more sensitive to the presence of thermal lenses. However, recently a first demonstration of this multi-pass concept in a thin-disk KLM oscillator was realized with six double passes through the thin disk per round-trip resulting in 130 W average and 20 MW peak output power [63]. With respect to a reference oscillator, the peak power did not drop when increasing the output-coupling ratio to 30%, rendering it an encouraging result towards scaling the output coupling ratio to 50%.

4.2. Positive dispersion regime

All oscillators presented (see Table 1) were mode locked in the anomalous dispersion regime providing bandwidth limited, unchirped pulses with a well behaved temporal phase. However, this implies high peak-intensities inside the oscillator cavity and therefore strong nonlinear effects even in air. Other limitations might arise due to damage thresholds of the intra-cavity optics because of the intra-cavity peak intensities approaching several 100 GW / cm^2 and peak fluences up to several 10 mJ / cm^2 or even higher during the pulse build-up phase. Up to now, these high intensities have not posed the major limitation to thin-disk KLM oscillators; however, this situation might change in the future when even higher peak and average powers will be targeted, especially in combination with a compact resonator design. Favorably low intensities can be provided by the pulse formation in the normal dispersion regime (chirped-pulse regime) which was first investigated in Ti:Sa oscillators [64] and is nowadays commonly employed to increase the pulse energies obtainable from fiber oscillators [65]. The pulses that form inside such an oscillator are strongly chirped, resulting in lower peak-powers at the same pulse-energy as compared to the solitons under anomalous dispersion. In contrast to Eq. (1), these pulses theoretically scale better in pulse energy with respect to the dispersion-compensation such that $E \propto \beta 2$ [66]. While in Ti:Sa oscillators, this method of mode-locking allowed a major improvement in pulse energy [44, 67], the output from Yb-based mode-locked thin-disk oscillators did not improve over the anomalous dispersion regime [32, 68]. One of the reasons is the relatively narrow emission bandwidth of Yb:YAG and the necessity to introduce an additional spectral filter into the oscillator cavity. This spectral filtering was not performed in the work [68] due to additional complexity, losses and the high intra-cavity average power usually associated with thin-disk lasers. Moreover, no practical demand for the realization of a stable chirped-pulse regime has arisen till today since the limits of the anomalous dispersion regime in mode-locked thin-disk oscillators are not yet explored. However, this situation was different for the Ti:Sa bulk oscillators. Although this chirped-pulse regime appears attractive for power scaling and energy scaling [69], the downside seems an increased demand on the self-amplitude modulation to keep these pulses stable and provide reliable pulse-build-up. Due to the difficulties associated with this reliable pulse-build-up, the positive dispersion regime was not further investigated in thin-disk oscillators. To date the highest peak-powers are obtained from solitonic oscillators working in the anomalous dispersion regime and this situation is unlikely to change until some technical limitations associated with high intra-cavity average power and extremely low repetition rate (very long resonator length) will approached.

5. CONCISE SUMMARY

Ultimately, the further increase in average power of Kerr-lens mode-locked thin-disk oscillators will mostly be hindered by technological constraints associated with the thermal lensing in the dispersive mirrors. This issue can be overcome by utilizing larger beam diameters on the existing dispersive optics and using substrates with high thermal conductivity. The TEM00 continuous wave (CW) performance of thin-disk lasers can be considered as the maximum limit for average power scaling, which now exceeds 1 kW [70].

The current challenge in scaling the peak power of this technology is the impact of lower air pressure. However, this issue may be largely overcome by employing the active multi-pass method and increasing the output coupling ratios. Nevertheless, this constraint is mostly of a technical nature and does not impose a fundamental restriction on achieving output peak powers in the gigawatt range. The combination of the geometrical energy scaling concept and the intrinsic advantages of the KLM approach allows for peak power scalability. However, this comes at the cost of reduced ambient pressure. A more basic constraint arises from the challenges associated with initiating mode-locking. Put simply, a mode-locking element must be capable of initiating oscillation with very low peak power and maintaining stable mode-locked functioning even at peak values above 10 GW within the cavity. This means it needs to cover a wide range of peak powers. The requirements for initiating and maintaining stable operation are inherently contradictory, making this constraint present in all forms of mode-locked oscillators.

AT A GLANCE

Femtosecond Kerr-lens mode-locked thin-disk oscillators are a type of oscillator that can be scaled in terms of both peak and average power. In recent years, there has been significant progress in the development of amplification systems that can achieve exceptionally high average and peak power levels. These systems can now deliver power levels exceeding 200 W and 50 MW, respectively, which is beyond the capabilities of more traditional amplification systems. These advancements were followed by numerous difficulties, such as the implementation of mode-locking, thermal lensing, and maintaining oscillator stability. These obstacles were effectively surmounted, leading to an enhanced comprehension of the power scaling of this technology. We provide a comprehensive analysis of these many features and demonstrate that this technology has a highly promising future, not only in terms of further power scaling but also in terms of its applications. Specifically, this oscillator has the capability to create a unique category of small, portable, high-intensity extreme-ultraviolet and infrared radiation sources, which will open up new possibilities for spectroscopic applications.

REFERENCES

1. Russbueldt P, Hoffmann D, Hofer M, Lohring J, Luttmann J, Meissner A, et al. Innoslab amplifiers. IEEE Journal of Selected Topics in Quantum Electronics. 2015;21(1):447-463. DOI: 10.1109/JSTQE.2014.2333234
2. Eidam T, Hanf S, Seise E, Andersen TV, Gabler T, Wirth C, et al. Femtosecond fiber CPA system emitting 830 W average output power. Optics Letters. 2010;35(2):94-96. DOI: 10.1364/OL.35.000094
3. Fattahi H, Barros HG, Gorjan M, Nubbemeyer T, Alsaif B, Teisset CY, et al. Third-generation femtosecond technology. Optica. 2014;1(1):45-63. DOI: 10.1364/OPTICA.1.000045
4. Zhang J, Brons J, Lilienfein N, Fedulova E, Pervak V, Bauer D, et al. 260-megahertz, megawatt-level thin-disk oscillator. Optics Letters. 2015;40(8):1627-1630. DOI: 10.1364/OL.40.001627
5. Pronin O, Seidel M, Lucking F, Brons J, Fedulova E, Trubetskov M, et al. High-power multi-megahertz source of waveform-stabilized

few-cycle light. Nature Communications. 2015;6:6988. DOI: 10.1038/ncomms7988

6. Seidel M, Xiao X, Hussain SA, Arisholm G, Hartung A, Zawilski KT, et al. Multi-watt, multi-octave, mid-infrared femtosecond source. Science Advances. 2018;4(4):1526

7. Brons J, Pervak V, Fedulova E, Bauer D, Sutter D, Kalashnikov V, et al. Energy scaling of Kerr-lens mode-locked thin-disk oscillators. Optics Letters. 2014;39(22):6442-6445. DOI: 10.1364/OL.39.006442

8. Brons J, Pervak V, Bauer D, Sutter D, Pronin O, Krausz F. Powerful 100-fs-scale Kerr-lens mode-locked thin-disk oscillator. Optics Letters. 2016;41(15):3567-3570. DOI: 10.1364/OL.41.003567

9. Zhang J, Brons J, Seidel M, Bauer D, Sutter D, Pervak V, et al., editors. Generation of 49-fs Pulses Directly from Distributed Kerr-lens Mode-locked Yb:YAG Thin-disk Oscillator. Advanced Solid State Lasers. Berlin: OSA; 2015

10. Seidel M, Brons J, Arisholm G, Fritsch K, Pervak V, Pronin O. Efficient high-power Ultrashort pulse compression in self-defocusing bulk media. Scientific Reports. 2017;7(1):1410. DOI: 10.1038/s41598-017-01504-x

11. Bechtel HA, Muller EA, Olmon RL, Martin MC, Raschke MB. Ultrabroadband infrared nanospectroscopic imaging. Proceedings of the National Academy of Sciences. 2014;111(20):7191-7196. DOI: 10.1073/pnas.1400502111

12. de Oliveira N, Roudjane M, Joyeux D, Phalippou D, Rodier J-C, Nahon L. High-resolution broad-bandwidth Fourier-transform absorption spectroscopy in the VUV range down to 40 nm. Nature Photon. 2011;5:149. DOI: 10.1038/nphoton.2010.314

13. Udem T, Holzwarth R, Hansch TW. Optical frequency metrology. Nature. 2002;416(6877):233-237. DOI: 10.1038/416233a

14. Newbury NR. Searching for applications with a fine-tooth comb. Nature Photon. 2011;5(4):186-188

15. Hentschel M, Kienberger R, Spielmann C, Reider GA, Milosevic N, Brabec T, et al. Attosecond metrology. Nature. 2001;414(6863):509-513. DOI: 10.1038/35107000

16. Krebs M, Hadrich S, Demmler S, Rothhardt J, Zair A, Chipperfield L, et al. Towards isolated attosecond pulses at megahertz repetition rates. Nature Photon. 2013;7(7):555-559. DOI: 10.1038/nphoton.2013.131

17. Köhler B, Kissel H, Flament M, Wolf P, Brand T, Biesenbach J, editors. High-power diode laser modules from 410 nm to 2200 nm. Proceedings of SPIE. Vol. 7583. 2010

18. Spence DE, Kean PN, Sibbett W. 60-fsec pulse generation from a self-mode-locked Ti:sapphire laser. Optics Letters. 1991;16(1):42-44

19. Moulton PF. Spectroscopic and laser characteristics of Ti:Al_2O_3. Journal of the Optical Society of America B: Optical Physics. 1986;3(1):125-133. DOI: 10.1364/JOSAB.3.000125

20. Steffen H, Jan R, Manuel K, Stefan D, Arno K, Andreas T, et al. Single-pass high harmonic generation at high repetition rate and photon flux. Journal of Physics B: Atomic, Molecular and Optical Physics. 2016;49(17):172002

21. Pupeza I, Sánchez D, Zhang J, Lilienfein N, Seidel M, Karpowicz N, et al. High-power sub-two-cycle mid-infrared pulses at 100 MHz repetition rate. Nature Photon. 2015;9(11):721-724. DOI: 10.1038/nphoton.2015.179

22. Gohle C, Udem T, Herrmann M, Rauschenberger J, Holzwarth R, Schuessler HA, et al. A frequency comb in the extreme ultraviolet. Nature. 2005;436(7048):234-237. DOI: 10.1038/nature03851

23. Benko C, Allison TK, Cingoz A, Hua L, Labaye F, Yost DC, et al. Extreme ultraviolet radiation with coherence time greater than 1 s. Nature Photon. 2014;8(7):530-536. DOI: 10.1038/nphoton.2014.132

24. Carstens H, Högner M, Saule T, Holzberger S, Lilienfein N, Guggenmos A, et al. High-harmonic generation at 250 MHz with photon energies exceeding 100 eV. Optica. 2016;3(4):366-369. DOI: 10.1364/OPTICA.3.000366

25. Cingoz A, Yost DC, Allison TK, Ruehl A, Fermann ME, Hartl I, et al. Direct frequency comb spectroscopy in the extreme ultraviolet. Nature. 2012;482(7383):68-71. DOI: 10.1038/nature10711

26. Brons J. High-power femtosecond laser-oscillators for applications in high-field physics [Dissertation]. LMU Munich; 2017

27. Aus der Au J, Spühler GJ, Südmeyer T, Paschotta R, Hövel R, Moser M, et al. 16.2-W average power from a diode-pumped femtosecond Yb:YAG thin disk laser. Optics Letters. 2000;25(11):859-861. DOI: 10.1364/OL.25.000859

28. Saraceno CJ, Emaury F, Schriber C, Hoffmann M, Golling M, Sudmeyer T, et al. Ultrafast thin-disk laser with 80 μJ pulse energy and 242 W of average power. Optics Letters. 2014;39(1):9-12. DOI: 10.1364/OL.39.000009

29. Saraceno CJ, Emaury F, Heckl OH, Baer CRE, Hoffmann M, Schriber C, et al. 275 W average output power from a femtosecond thin disk oscillator operated in a vacuum environment. Optics Express. 2012;20(21):23535-23541. DOI: 10.1364/OE.20.023535

30. Bauer D, Zawischa I, Sutter DH, Killi A, Dekorsy T. Mode-locked Yb:YAG thin-disk oscillator with 41 μJ pulse energy at 145 W average infrared power and high power frequency conversion. Optics Express. 2012;20(9):9698-9704. DOI: 10.1364/OE.20.009698

31. Neuhaus J, Bauer D, Kleinbauer J, Killi A, Sutter DH, Dekorsy T. Numerical analysis of a sub-picosecond thin-disk laser oscillator with active multipass geometry showing a variation of pulse duration within one round trip. Journal of the Optical Society of America B: Optical Physics. 2010;27(1):65-71. DOI: 10.1364/JOSAB.27.000065

32. Palmer G, Schultze M, Siegel M, Emons M, Bunting U, Morgner U. Passively mode-locked Yb:KLu$(WO_4)_2$ thin-disk oscillator operated in the positive and negative dispersion regime. Optics Letters. 2008;33(14):1608-1610

33. Pronin O, Brons J, Grasse C, Pervak V, Boehm G, Amann MC, et al. High-power 200 fs Kerr-lens mode-locked Yb:YAG thin-disk oscillator. Optics Letters. 2011;36(24):4746-4748. DOI: 10.1364/OL.36.004746

34. Wintner E, Sorokin E, Sorokina I, inventors. Laser system for producing ultrashort laser pulses. 2002

35. Available from: https://www.rp-photonics.com/spotlight_2011_12_23.html
36. Eilanlou AA, Nabekawa Y, Kuwata-Gonokami M, Midorikawa K. Femtosecond laser pulses in a Kerr lens mode-locked thin-disk ring oscillator with an intra-cavity peak power beyond 100 MW. Japanese Journal of Applied Physics. 2014;53(8):082701. DOI: 10.7567/jjap.53.082701
37. Paradis C, Modsching N, Wittwer VJ, Deppe B, Kränkel C, Südmeyer T. Generation of 35-fs pulses from a Kerr lens mode-locked Yb:Lu_2O_3 thin-disk laser. Optics Express. 2017;25(13):14918-14925. DOI: 10.1364/OE.25.014918
38. Kreipe B, Andrade JRCd, Kränkel C, Morgner U, editors. Kerr-lens Mode-locked Yb:Lu_2O_3 Thin-disk Laser. 2016 Conference on Lasers and Electro-Optics (CLEO); 2016. pp. 5-10
39. Piche M. Beam reshaping and self-mode-locking in nonlinear laser resonators. Optics Communication. 1991;86(2):156-160. DOI: 10.1016/0030-4018(91)90552-O
40. Lariontsev EG, Vladimir NS. Possibility of using self-focusing for increasing contrast and narrowing of ultrashort light pulses. Soviet Journal of Quantum Electronics. 1975;5(7):796
41. Marconi MC, Martinez OE, Diodati FP. Short pulse generation in solid state lasers by a novel passive technique. Optics Communication. 1987;63(3):211-216. DOI: 10.1016/0030-4018(87)90159-3
42. Haus HA, Fujimoto JG, Ippen EP. Structures for additive pulse mode locking. Journal of the Optical Society of America B: Optical Physics. 1991;8(10):2068-2076. DOI: 10.1364/JOSAB.8.002068
43. Mollenauer LF, Stolen RH. The soliton laser. Optics Letters. 1984;9(1):13-15. DOI: 10.1364/OL.9.000013
44. Naumov S, Fernandez A, Graf R, Dombi P, Krausz F, Apolonski A. Approaching the microjoule frontier with femtosecond laser oscillators. New Journal of Physics. 2005;7(1):216
45. Saraceno CJ, Emaury F, Schriber C, Diebold A, Hoffmann M, Golling M, et al. Toward Millijoule-level high-power ultrafast thin-disk

oscillators. Selected Topics in Quantum Electronics, IEEE Journal of. 2015;21(1):1-18. DOI: 10.1109/JSTQE.2014.2341588

46. Huonker M, Schmitz C, Voss A. Laser amplifying system. Google Patents; 2005

47. Dominik B. Modengekoppelte Scheibenoszillatoren für Materialbearbeitung: University of Konstanz; 2012

48. Georgiev D, Herrmann J, Stamm U. Cavity design for optimum nonlinear absorption in Kerr-lens mode-locked solid-state lasers. Optics Communication. 1992;92(4):368-375. DOI: 10.1016/0030-4018(92)90647-A

49. Brabec T, Curley PF, Spielmann C, Wintner E, Schmidt AJ. Hard-aperture Kerr-lens mode locking. Journal of the Optical Society of America B: Optical Physics. 1993;10(6):1029-1034. DOI: 10.1364/JOSAB.10.001029

50. Weber MJ. Handbook of Optical Materials. Boca Raton, FL; London: CRC Press; 2003. 512 p

51. Pronin O. Towards a Compact Thin-Disk-Based Femtosecond XUV Source. Springer; 2013. 144 p

52. Fedulova E. Development of dielectric multilayer coatings for ultrafast oscillators [dissertation]. LMU Munich; 2016

53. Pervak V, Pronin O, Razskazovskaya O, Brons J, Angelov IB, Trubetskov MK, et al. High-dispersive mirrors for high power applications. Optics Express. 2012;20(4):4503-4508. DOI: 10.1364/OE.20.004503

54. Carstens H, Lilienfein N, Holzberger S, Jocher C, Eidam T, Limpert J, et al. Megawatt-scale average-power ultrashort pulses in an enhancement cavity. Optics Letters. 2014;39(9):2595-2598. DOI: 10.1364/OL.39.002595

55. Hädrich S, Rothhardt J, Demmler S, Tschernajew M, Hoffmann A, Krebs M, et al. Scalability of components for kW-level average power few-cycle lasers. Applied Optics. 2016;55(7):1636-1640. DOI: 10.1364/AO.55.001636

56. Agrawal GP. Chapter 5—Optical Solitons. In: Nonlinear Fiber Optics. 5th ed. Boston: Academic Press; 2013. pp. 129-191

57. Fedulova E, Fritsch K, Brons J, Pronin O, Amotchkina T, Trubetskov M, et al. Highly-dispersive mirrors reach new levels of dispersion. Optics Express. 2015;23(11):13788-93. DOI: 10.1364/OE.23.013788

58. Pronin O, Seidel M, Brons J, Lücking F, Angelov IB, Kalashnikov VL, et al., editors. Towards CEP stabilized pulses from a KLM Yb:YAG thin-disk oscillator. In: 2013 Conference on Lasers & Electro-Optics Europe & International Quantum Electronics Conference CLEO EUROPE/IQEC; 2013 12–16. May 2013

59. Klenner A, Emaury F, Schriber C, Diebold A, Saraceno CJ, Schilt S, et al. Phase-stabilization of the carrier-envelope-offset frequency of a SESAM modelocked thin disk laser. Optics Express. 2013;21(21):24770-24780. DOI: 10.1364/OE.21.024770

60. Seidel M, Brons J, Lücking F, Pervak V, Apolonski A, Udem T, et al. Carrier-envelope-phase stabilization via dual wavelength pumping. Optics Letters. 2016;41(8):1853-1856. DOI: 10.1364/OL.41.001853

61. Gröbmeyer S, Brons J, Seidel M, Krausz F, Pronin O, editors. 100 W-level carrier-envelope-phase stable thin-disk oscillator. XXIst International Conference on Ultrafast Phenomena; 2018; Hamburg

62. Neuhaus J, Bauer D, Zhang J, Killi A, Kleinbauer J, Kumkar M, et al. Subpicosecond thin-disk laser oscillator with pulse energies of up to 25.9 microjoules by use of an active multipass geometry. Optics Express. 2008;16(25):20530-20539. DOI: 10.1364/OE.16.020530

63. Poetzlberger M, Brons J, Zhang J, Bauer D, Sutter D, Krausz F, et al., editors. Towards Active Multipass Kerr-lens Mode-locked Yb:YAG Thin-disk Oscillators. 2017 European Conference on Lasers and Electro-Optics and European Quantum Electronics Conference; 2017; Munich: Optical Society of America

64. Proctor B, Westwig E, Wise F. Characterization of a Kerr-lens mode-locked Ti:sapphire laser with positive group-velocity dispersion. Optics Letters. 1993;18(19):1654-1656. DOI: 10.1364/OL.18.001654

65. Ilday FÖ, Buckley JR, Wise FW, editors. Self-similar evolution of parabolic pulses in a fiber laser. Nonlinear Guided Waves and Their Applications. Toronto: Optical Society of America; 2004

66. Kalashnikov VL. Chirped-pulse oscillators: Route to the energy-scalable femtosecond pulses. In: Amin Al-Khursan, editor. Solid State Laser. InTech; 2012. ISBN: 978-953-51-0086-7

67. Kalashnikov VL, Podivilov E, Chernykh A, Naumov S, Fernandez A, Graf R, et al. Approaching the microjoule frontier with femtosecond laser oscillators: Theory and comparison with experiment. New Journal of Physics. 2005;7(1):217

68. Pronin O, Brons J, Grasse C, Pervak V, Boehm G, Amann MC, et al. High-power Kerr-lens mode-locked Yb:YAG thin-disk oscillator in the positive dispersion regime. Optics Letters. 2012;37(17):3543-3545. DOI: 10.1364/OL.37.003543

69. 69.Renninger WH, Wise FW. Fundamental limits to mode-locked lasers: Toward terawatt peak powers. IEEE Journal of Selected Topics in Quantum Electronics. 2015;21(1):1-8. DOI: 10.1109/JSTQE.2014.2329936

70. Schad S-S, Gottwald T, Kuhn V, Ackermann M, Bauer D, Scharun M, et al., editors. Solid State Lasers XXV: Technology and Devices. In: Proceedings of SPIE. Clarkson WA, Shori RK, editor. 2016;9726:972615

CHAPTER-6

CREATING 2 μM TM3⁺ -DOPED FIBER LASERS WITH HIGH-ENERGY DISSIPATIVE SOLITONS

1. INTRODUCTION: AN OVERVIEW

Ultrafast 2 μm fiber lasers have gained significant global interest in recent years and have been widely used in various fields such as LIDAR, surgical procedures, molecule spectroscopy, optical sensing, medical treatment, material processing, and nonlinear microscopy [1, 2, 3, 4, 5, 6, 7, 8]. Ultrashort pulses in applications typically necessitate a high pulse energy, which holds significant importance for scientific and industrial purposes. Laser scientists persistently strive to achieve high-energy brief pulses at diverse wavelengths.

Fiber lasers (FLs) are superior to traditional solid-state lasers (SSLs) for generating ultrashort laser pulses because they are more compact, resilient, and produce high-quality laser beams. Traditionally, the soliton mode-locking technique is used to generate brief pulses in fiber networks. Several passive mode-locking approaches can be utilized, such as nonlinear polarization rotation (NPR) [9, 10], nonlinear loop mirror [11, 12], and saturable absorber method [13, 14]. Nevertheless, the pulse energy of conventional solitons, which rely on the equilibrium between dispersion and nonlinearity and have anomalous net cavity dispersion, is typically constrained to less than 1 nJ due to the soliton area theorem [15, 16] or the pulse peak power clamping effect [17, 18]. Thus, fiber lasers continue to generate significantly lower pulse energy compared to solid-state lasers [19].

Various techniques have been proposed and explored to enhance the pulse energy of fiber lasers. These techniques include dispersion-managed soliton, all normal dispersion mode-locking, self-similar soliton, and dissipative soliton. Mechanisms such as dispersion-managed soliton, all normal dispersion mode-locking, self-similar soliton, and dissipative soliton have been found to be particularly effective in improving pulse energy. These mechanisms have been extensively studied and their effectiveness has been demonstrated in multiple studies [20, 21, 22, 23, 24, 25, 26, 27, 28, 29, 30, 31, 32, 33, 34, 35, 36, 37, 38, 39]. DS mode-locked fiber lasers have achieved pulse energy that is 10 to 100 times more than traditional soliton mode-locking lasers by utilizing the balance of nonlinearity, dispersion, gain, and loss [31, 32]. Despite the fact that the pulse energy of 1 to 1.5 μm fiber lasers has surpassed 10 nJ [40, 41, 42] and even reached over 20 nJ [33, 34, 35], the pulse energy of 2 μm DS fiber lasers remains relatively low. The reason for this is that the gain fibers (GFs) now available in the 2 μm range exhibit significant anomalous dispersion. As a result, standard soliton mode-locking operation of 2 μm fiber lasers is achieved [43, 44, 45, 46, 47, 48]. Thus, the pulse energy is still determined by the soliton area theorem and restricted by peak power [15, 17].

The utilization of DS mode-locking has become prevalent as an effective technique for enhancing the pulse energy of 2 μm fiber lasers. In order to achieve DS mode-locking, it is necessary to shift the dispersion of the entire cavity into the normal dispersion zone. In order to achieve this objective, several techniques have been suggested, such as the inclusion of a chirped fiber Bragg grating in the cavity to introduce normal dispersion [49], or the integration of specifically engineered dispersion-compensating fibers (DCFs) into the cavity [50, 51, 52]. Nevertheless, these techniques merely enhance the pulse energy to approximately 1 nJ, leaving the vast potential of the DS mode-locking mechanism still unexplored.

In this study, we will introduce a novel model to examine the pulsing dynamics within a 2 μm DS mode-locked fiber laser. We will demonstrate that, unlike their 1 to 1.5 μm counterparts, the pulse energy of 2 μm DS fiber lasers is primarily constrained by the nonlinear phase shift resulting from the gain fiber. Consequently, we suggest that the anomalous dispersive gain fiber should be minimized in length to effectively separate the effects of gain, dispersion, and nonlinearity. The term we use to refer to it is condensed-gain fiber mode-locking (CGFML). Numerical simulations demonstrate that it is easily achievable to have over 10 nJ of dissipative solitons (DSs) at a wavelength of 2 μm by minimizing excessive phase buildup. Subsequently, we conducted experimental manipulation of a 2 μm fiber laser with CGFML. As a result, we achieved a 4.9 nJ DS with a dechirped pulse duration of 579 fs. By further enhancing the cavity, the pulse energy of this mode-locked fiber laser can be increased to more than 12 nJ, which is similar to that of traditional solid-state mode-locked lasers. The study focuses on analyzing the pulsing dynamics of the CGFML fiber laser under high pumping levels, leading to the successful achievement of high-energy harmonic mode-locking. Here, we demonstrate the successful implementation of high-energy mode-locking in a 2 μm fiber laser using newly discovered 2D materials (MoS2). Additionally, we have achieved a pulse energy of over 15 nJ. The research findings suggest that the CGFML offers a novel approach for the advancement of high-energy fiber lasers with a short pulse duration in the 2 μm wavelength range.

2. CONDENSED-GAIN DISSIPATIVE SOLITON MODEL AND SIMULATION FOR 2 MM FIBER LASERS

Based on detailed simulation of the dynamics of short pulse propagating in various fiber circumstances, we found that the main factor that limited the pulse energy in 2 μm DS fiber lasers was related with nonlinear phase shift, which was primarily accumulated in the gain fiber. If we can efficiently control the nonlinear phase shift generated in the gain fiber, then the pulse energy of 2 μm DS fiber lasers probably can be significantly scaled. Therefore, we propose a condensed-gain fiber model where the gain fiber should be shortened as much as possible, and in the following, we give a detailed description about the model and carry out simulation about the pulse dynamics happened in a 2 μm DS fiber laser.

A simple schematic diagram for the CGFML model is shown in Figure 1(a) [53]. The fiber laser cavity mainly includes five elements: output coupler (OC), single-mode fiber (SMF), gain fiber (GF), dispersion-compensating fiber (DCF), and saturable absorber (SA). Here, we use a single-mode highly doped 2 μm thulium fiber as the GF. Light evolution (pulse shape, pulse intensity, and spectrum) in the laser cavity is traced through solving the well-known nonlinear Schrodinger equation (NLSE) [27], which needs the original equation:

$$\frac{\partial U(z,\tau)}{\partial z} + i\frac{\beta_2}{2}\frac{\partial^2 U(z,\tau)}{\partial \tau^2} = i\gamma\left|U(z,\tau)\right|^2 U(z,\tau) + gU(z,\tau) \quad (1)$$

where U(z, τ) is the envelope of the light field, z is the propagation coordinate, and τ is the time-delay parameter. The SMF (8.2/125 μm, 0.14 NA) has a length of 1.4 m, with β_2 = -67 ps²/km and γ = 0.001 (Wm)$^{-1}$, while the DCF (2.2/125 μm, 0.35 NA) is 1.5 m long with β_2 = 93 ps²/km and γ = 0.007 (Wm)$^{-1}$, respectively. The 0.2 m GF (5/125 μm, 0.24 NA), with β_2 = - 12 ps²/km and γ = 0.003 (Wm)$^{-1}$, has the gain (including saturation) as.

$$g = g_0/\left[1 + E_{pulse}/E_{sat} + (\omega - \omega_0)^2/\Delta\omega^2\right] \quad (2)$$

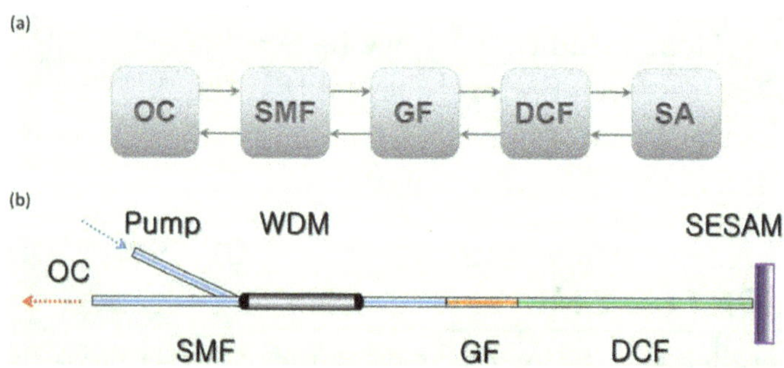

Figure 1. (a) Schematic diagram of the condensed-gain fiber laser model shows the light flow in the cavity. (b) Experimental setup of the SESAM mode-locked fiber laser system with a linear cavity. OC, output coupler; SMF, single-mode fiber; GF, gain fiber; DCF, dispersion-compensating fiber; SA, saturable absorber [53].

where g_0 is the small-signal gain (here it is taken to be 30 dB), E_{pulse} is the pulse energy, E_{sat} is the gain saturation energy, ω_0 is the gain-center angular frequency, and Δω is the gain bandwidth (assume 90 nm).

The saturable absorption effect of SA is included by using the transfer function:

$$T = 1 - l_0/\left[1 + P(\tau)/P_{sat}\right] \quad (3)$$

where l_0 is the unsaturated loss (take 0.7), P(τ) is the instantaneous power, and P_{sat} is the saturation power. A critical factor for achieving DS is that spectral filtering is required to balance gain and loss. To that end, a 150-nm-bandwidth spectral filter (SF) is exerted on the SA.

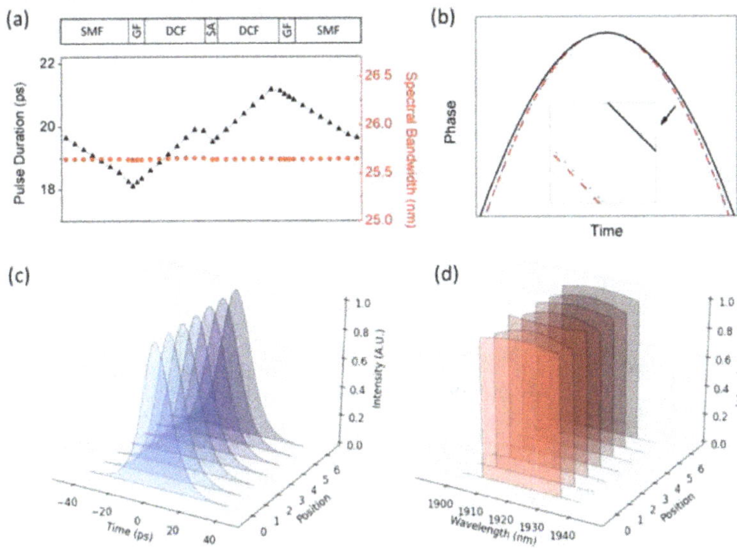

Figure 2. Evolution dynamics of pulse duration (black triangles) and spectral bandwidth (red circles) through different elements inside the laser cavity (a) and temporal phase of the pulse after DCF (black solid), GF (red dashed), and SMF (blue dotted) (b) and temporal (c) and spectral profiles (d) of the pulse after successive elements [53].

Here, the split-step Fourier method is used to solve the NLSE. Simulation is started as the following procedure. With an initial white noise, the light is calculated in both temporal and spatial regions until a steady state is reached. The pump level and saturable effect are controlled through changing the values of E_{sat} and P_{sat}. When we take P_{sat} = 3.5 kW and E_{sat} = 3.4 nJ, the pulse's temporal and spectral evolution characteristics are shown in Figure 2 [53]. In this case, the stable solution of pulse energy is 5 nJ. Detailed variations of pulse shape/width and spectral shape are clearly shown. In the DCF, owing to the combined effects of normal group velocity dispersion (GVD) and nonlinearity (NL), the pulse propagates with its duration increasing monotonically. The broadened pulse is then compressed by the SMF and GF with anomalous GVD. The pulse's spectrum has steep edges, and the bandwidth negligibly changes during the pulse circulating inside the cavity. However, the spectrum shape shows characteristic changes during the pulse evolution. The GF tends to amplify the spectrum's center more, and thus, the spectrum top becomes more arched. At the same time, the amplified pulse gives rise to increased self-phase modulation, thus leading to sharp edge peaks of the spectrum. Then, after being successively shaped by SA, DCF, GF, and SMF, the spectrum recovers its nearly flat-top shape. We can also see the advantages of the condensed-gain fiber model from the phase shift during the pulse evolution. As shown in Figure 2(b) [53], after passing through the three different kinds of fibers (DCF, GF, and SMF), very little phase shift is accumulated by the GF, which is finally compensated by both the DCF and the SMF.

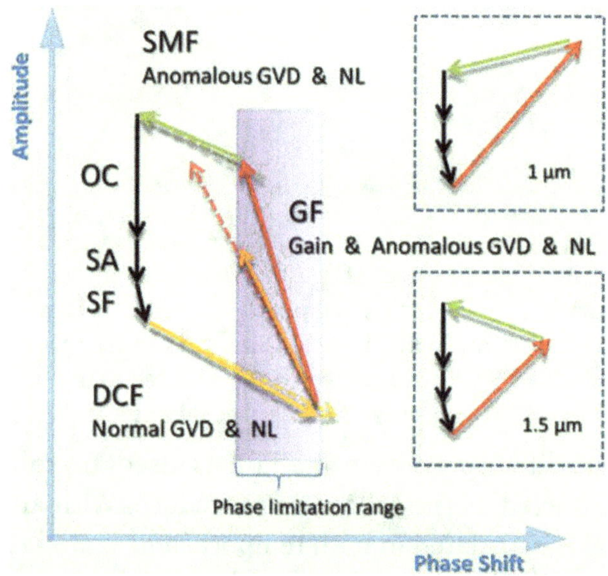

Figure 3. Schematic diagram for the amplitude and phase balances in a 2 μm fiber laser (insets show the counterparts of 1 μm and 1.5 μm systems) [53].

To gain deeper insight into the intracavity pulsing dynamics, a qualitative illustration for 2 μm DSs is summarized in Figure 3 [53], along with their 1 and 1.5 μm counterparts (insets) [54]. As shown in the insets, in the case of 1 and 1.5 μm, the GFs (Yb-doped or Er-doped) have normal dispersion and introduce positive phase shift, which can be compensated (even if the shift is large) by the negative phase shift provided by the SMF. However, it is quite different in the 2 μm wavelength regime, where the GF (Tm-doped) has anomalous dispersion and, thus, negative phase shift. To achieve soliton mode-locking, normally dispersive fibers (DCF and SMF) are thus required to be integrated into the cavity. However, too large normal dispersion value (long fibers) will induce large phase shift and consequently pulse splitting. Therefore, small net normal dispersion (caused both by DCF and SMF) places a tolerant phase shift region for the GF (purple area). A longer GF will induce much more significant phase shift (red-dashed arrow) than that incurred by the DCF or SMF (yellow- or green-dashed arrow). In the phase limitation range, shorter GF (red arrow) has a larger slope and hence can achieve higher pulse energy. On the contrary, longer GF (orange arrow), due to its smaller slope, has to sacrifice a large part of amplitude to reduce its phase shift under the tolerable level. Therefore, short GF should be adopted to achieve high energy pulses from a cavity in the 2 μm spectral region.

Based on the above analysis, we propose the condensed GF (shortened to a small length while providing adequate gain at the same time) to scale the pulse energy of DSs in the 2 μm and mid-infrared spectral regions. Within the phase limitation range, a condensed GF has a large slope (Figure 3 [53]) and thereby can provide high pulse energy.

To verify the advantages of CGFML in the 2 μm regime, we carry out simulations in a semiconductor saturable absorber mirror (SESAM) mode-locked fiber laser (Figure 1 [53]). The simulated maximum output pulse energies with various GF lengths are indicated in Figure 4(a) [53]. It is clear that decreasing the GF length will dramatically increase the pulse energy. Shortening the GF to 0.2 m, as high as 11 nJ pulses, is achieved, which is much higher than the pulse energy of tradition solitons (usually less than 1 nJ). This thus confirms that CGFML is an effective route for generating high-energy soliton pulses in laser systems with anomalous dispersion GFs (shortening the anomalous dispersion GF as much as possible).

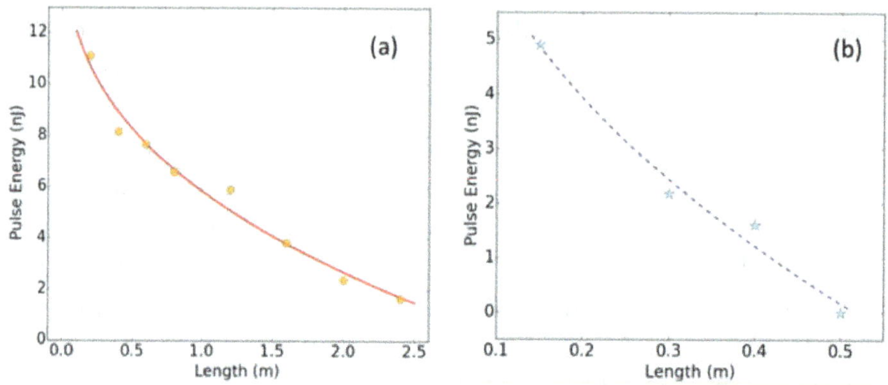

Figure 4. Simulated (circle dots) (a) and measured (asterisk dots) (b) maximum pulse energy under different lengths of GF under the pump power of 1 W. The curves are exponential fittings [53].

3. HIGH-ENERGY DISSIPATIVE SOLITON 2 MM FIBER LASERS

To verify the simulation results, we carried out a corresponding experimental observation of 2 μm DS mode-locking with short GFs. These GFs are highly thulium doped, and the net cavity dispersion is kept normal through adjusting the length of the DCF and SMF. Experimental setup is schematically shown in Figure 1(b) [53]. The pump source is a 1550-nm-CW Er/Yb-codoped fiber laser with maximum output of 1 W. The pump light is delivered into the gain fiber (with absorption of ~1.2 dB/cm at 1550 nm) by wavelength-division-multiplexing (WDM) couplers with a coupling efficiency of 95%. The parameters of these three kinds of fibers are the same as used in the simulation. Total cavity dispersion is kept at a net normal value of ~0.04 ps^2. On the output end, the perpendicularly cleaved fiber facet is employed for both laser feedback (~4% Fresnel reflection) and the output coupler (~96% output coupling ratio). The right side fiber end is directly butt coupled to the SESAM with a reflectance of ~85% at 1900 nm, a modulation depth of ~25%, and a saturation fluence of ~35 mJ/cm^2.

Short lengths of single-cladding Tm-doped fiber (tens of centimeters) are chosen as the GF. Under stable mode-locking operation, maximum pulse energies with different GF lengths are shown in Figure 4(b) [53]. The experimental results clearly follow the trend of the simulation prediction; that is, the pulse energy increases quickly as the length of GF decreases. When the GF length is shortened to ~15 cm, pulse energy of ~5 nJ is achieved, and detailed laser characteristics are shown in the following.

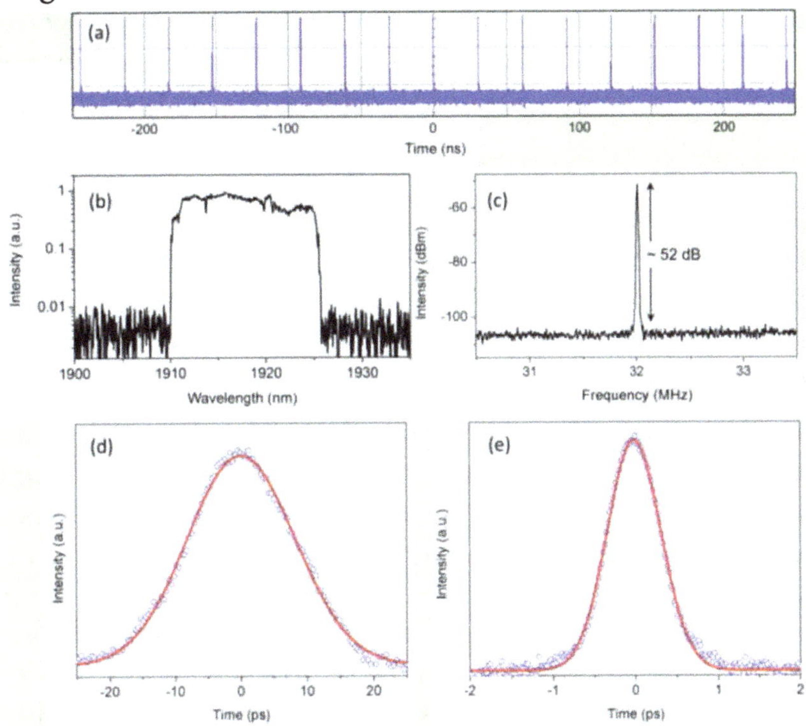

Figure 5. The laser pulse train (a), laser spectrum (b), and RF spectrum (c) of the mode-locked Tm-doped fiber laser and the autocorrelation traces of the pulse before (d) and after (e) dispersion compensation [53].

With the 15 cm GF, stable CW mode-locking is self-started when pump power is increased to ~650 mW. Owing to the large output coupling ratio suppressing the intermediate transitions between the CW laser operation and the CW mode-locking regime [55], no Q-switching or Q-switched mode-locking is observed. The stable CW mode-locked operation maintains when pump power is increased up to the maximum 1 W available pump power. The maximum average output power of this 2 μm DS fiber laser is 158 mW. Figure 5(a) [53] shows the 2 μm DS pulse train at the maximum output. The repetition rate is ~32 MHz, giving a pulse energy of ~4.9 nJ.

The laser spectrum, detected with a spectrometer (0.1 nm resolution), is shown in Figure 5(b). The center wavelength is 1918 nm and the 3 dB bandwidth is 15 nm. Steep spectral edges indicate the typical characteristics of DSs [31, 32]. The radio-frequency (RF) spectrum (Figure 5(c)) has a signal-to-noise ratio of ~52 dB, showing that the mode-locking state is very stable. We also use an autocorrelator to measure the pulse characteristics at the maximum output, and the pulse shape (autocorrelation (AC) trace) directly outputted from the laser cavity is indicated in Figure 5(d). The autocorrelation trace is fitted well by a Gaussian curve, giving a pulse duration of 16 ps. Therefore, the time-bandwidth product of the 2 μm DS pulse is calculated to be 18, which is highly chirped. For compressing this chirped pulse, we couple the output pulse directly into a ~25 m length of SMF-28 fiber. After dispersion compensation, the pulse is compressed to 579 fs (Figure 5(e)), and the time-bandwidth product reduces to 0.7.

This CGFML model can be readily extended to beyond 2 μm, e.g., mid-infrared fiber lasers to scale DS energy. According to this model, to achieve high-energy DSs, the GF length should be as short as possible to keep phase shift within the phase limitation range while providing enough gain at the same time. To this end, the GF should be highly doped (short length with enough gain), so that gain can decouple from dispersion and suppress phase shift. This can efficiently avoid the pulses' evolution to conventional solitons during amplification in the GF. This CGFML model can also be applied for ring laser cavities. In a ring cavity, the pulse passes every element only once during one cycle, leading to less phase shift accumulation. Therefore, a ring cavity has the potential to accommodate a larger phase limitation range and thus has a higher pulse energy generating possibility.

Scaling pulse energy. Based on the above condensed-gain model [53], we went to probe the upper limit of the pulse energy of 2 μm DS fiber lasers. In order to increase the pulse energy, we slightly increased the cavity fiber length, and therefore the pulse repetition rate will decrease and the pulse energy will be enhanced. In addition, we optimize the cavity parameters and manage the intracavity dispersion to scale the pulse energy of DS fiber laser in the 2 μm region. Here, we shorten the gain fiber to an optimal length and at the same time use a short piece of optimized single-mode fiber to compensate the dispersion.

The experimental system is illustrated in Figure 6 [56]. Pump light from a continuous-wave (CW) Er/Yb-codoped fiber laser with maximum output of ~1 W centered at 1550 nm was coupled into the fiber through a 1550/1900 nm WDM. The laser cavity includes two pieces of SMF-28 fibers, 12 cm length of single-mode Tm-doped silica fiber (5 μm core and 0.24 NA) and 1.5 m DCF. The DCF was butt coupled to the SESAM, whose reflection combined with the ~4% Fresnel reflection of the perpendicularly cleaved output fiber end completed the laser cavity. The dispersions of the thulium fiber, the SMF-28 fiber, and the DCF at 1920 nm were—12, -67, and 93 ps^2/km, respectively [52], giving a total net cavity dispersion of ~0.004 ps^2. The SESAM had a relaxation time of 10 ps and a modulation depth of 25%.

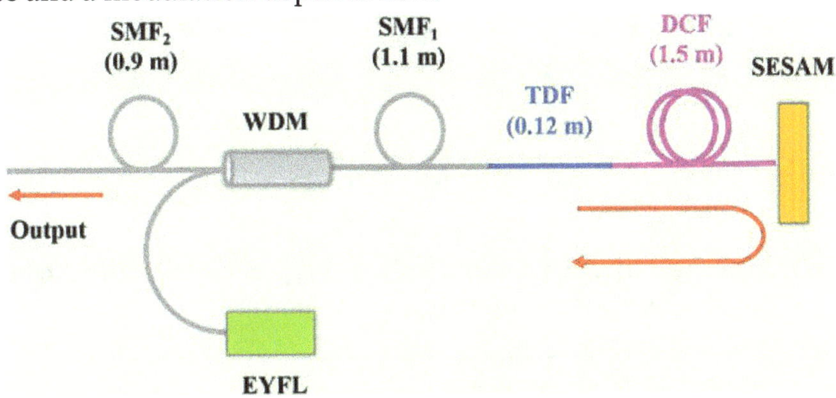

Figure 6. Schematic of the thulium-doped fiber laser passively mode-locked by a semiconductor saturable absorber mirror (SESAM). EYFL, erbium/ytterbium-codoped fiber laser; WDM, wavelength division multiplexer; SMF, single-mode fiber; TDF, thulium-doped fiber; DCF, dispersion-compensating fiber [56].

When we increased the pump power to 456 mW and at the same time carefully adjusted the SESAM, stable mode-locking was observed and could maintain up to the maximum pump power (1.09 W). Under different pump levels, the average output power was measured, and the pulse energy could be estimated based on the pulsing repetition rate. As shown in Figure 7, both average power and pulse energy increase near linearly with pump power, and the maximum output power and pulse energy are 263 mW and 12.07 nJ, respectively. Figure 8 shows the pulse duration versus pump power measured with an autocorrelator. The pulse duration displays a linear increase with pump power, indicating that the pulse was highly chirped. Large chirp is a typical characteristic of DSs for supporting high pulse energy. The laser spectrum, as shown in Figure 9 [56], locates at 1928.2 nm and has FWHM (full width at half maximum) bandwidth of 2.65 nm. The comparatively narrow spectrum width can be attributed to the high chirp-induced decrease of the pulse peak power. The spectrum shape is very similar to that of another recent report about DS fiber laser at the 1 μm regime [57].

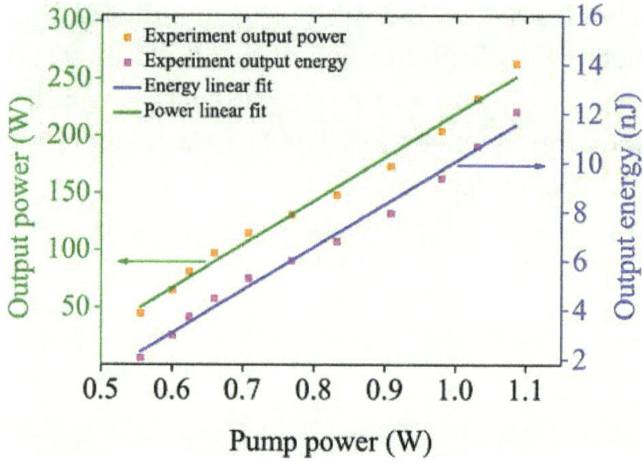

Figure 7. Average output power and pulse energy of the mode-locked fiber laser versus launched pump power [56].

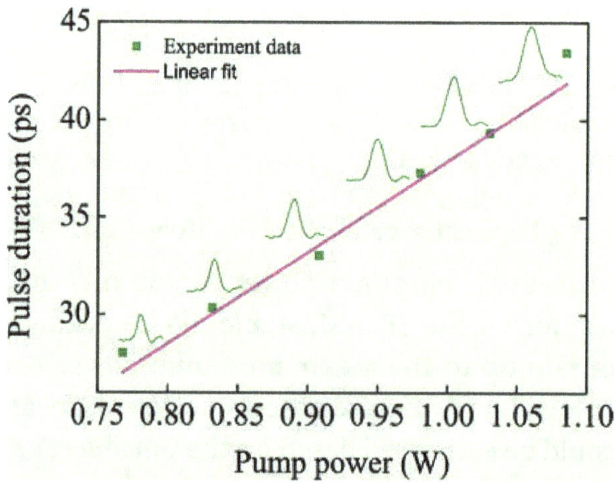

Figure 8. Pulse duration (autocorrelated trace) of the mode-locked fiber laser Versus launched pump power [56].

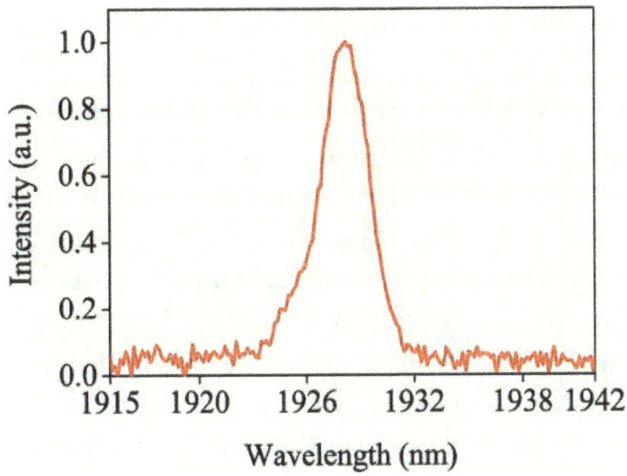

Figure 9. Laser spectrum of the mode-locked pulse [56].

Figure 10(a) shows the pulse train of the mode-locked fiber laser measured at the maximum output level. The pulse train has repetition rate of ~21.8 MHz, consistent with the total cavity length of 4.7 m. Figure 10(b) displays the autocorrelation trace of the pulse, giving a FWHM width of 43.5 ps if $sech^2$ pulse shape is assumed. Based on the spectral width of 2.65 nm, the pulse has a time-bandwidth product (TBWP) of ~9.3, indicating the presence of large chirp.

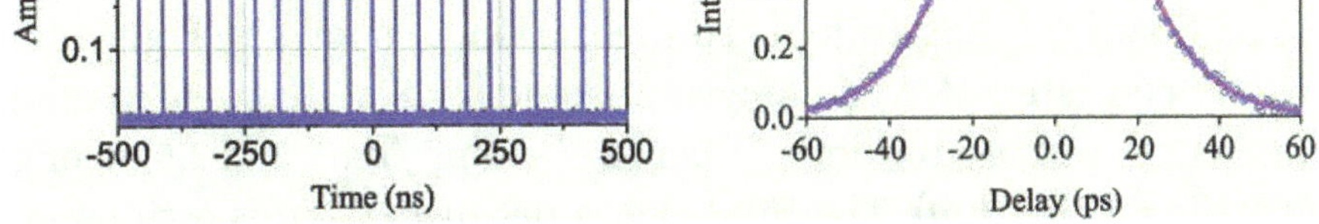

Figure 10. Dissipative soliton: (a) pulse train on oscilloscope and (b) autocorrelation trace of the single pulse at the maximum output level [56].

4. DISSIPATIVE SOLITON DYNAMICS OF 2 MM FIBER LASERS

The repetition rate of a passively mode-locked fiber laser is usually limited by the total cavity fiber length, and the pulsing repetition rate is generally of several MHz to tens of MHz. However, high-repetition-rate laser pulses are required in some application areas, including biological imaging [58], optical communication [59], and so on. If the high-repetition-rate laser pulses also have high pulse energy, then they are more preferred [60]. There are many ways to generate high-repetition-rate laser pulses from fiber lasers, but the most efficient one may be passive harmonic mode-locking. With harmonic mode-locking, the pulsing frequency will be highly multiplied just through increasing the intracavity light intensity to get higher-order harmonics. However, the single pulse energy usually decreases with increasing harmonic order. The pulse energy of harmonically mode-locked fiber lasers is limited by either pulsing instability or energy storage capability of fibers [61, 62]. In the 2 μm region, passively harmonic mode-locked fiber lasers, especially high-pulse-energy ones, are seldom reported.

Here, based on the CGFML and through appropriate designing the cavity dispersion map and adjusting the cavity gain, we experimentally realize multiple orders of harmonic mode-locking of 2 μm Tm-doped fiber laser (TDFLs) with a SESAM. To achieve high pulse energy, we design this laser to operate in the DS state and adopt a linear laser cavity. We observe stable harmonic mode-locking up to the fourth order, and the pulse energy of all these harmonic pulses is larger than 3 nJ, with the highest one being 12.37 nJ of the fundamental frequency pulsing. Besides harmonic mode-locking, we also observe soliton molecule mode-locking state of this 2 μm DS mode-locked fiber laser.

The laser system we adopted for the passively harmonic mode-locked 2 μm DS fiber laser has a simple configuration, as shown in Figure 11. It is mainly consisted of 1.1 m length of standard SMF (β_2 = -67 ps^2/km), 0.11 m length of thulium-doped fiber (β_2 = -12 ps^2/km), and 3.5 m length of DCF (β_2 = 93 ps^2/km). The total cavity net dispersion is estimated to be ~ 0.25 ps^2. A commercial 2 μm SESAM was adopted as the modulator, which has a modulation depth of 25% and relaxation time of 10 ps. A 1.1 W 1550 nm CW Er/Yb-codoped fiber laser was used as the pump source, and a WDM was adopted to launch the pump light into the cavity. The DCF was directly butt coupled to the SESAM. A 0.3-m-long SMF with one end perpendicularly cleaved was employed as the output coupler, and the ~4% fiber facet Fresnel reflection finishes the laser cavity together with the SESAM.

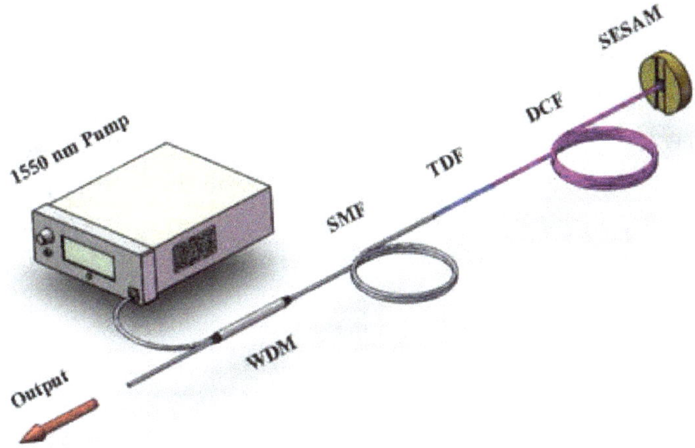

Figure 11. Schematic of the passively mode-locked thulium-doped fiber laser. WDM, wavelength division multiplexer; SMF, single-mode fiber; TDF, thulium-doped fiber; DCF, dispersion-compensating fiber; SESAM, semiconductor saturable absorber mirror [63].

First, we stimulated the laser to operate in the fundamental frequency mode-locking state. This was achieved through increasing the pump power to over a threshold value (here is 456 mW) and at the same time carefully adjusting the position of the SESAM. Once attained, the fundamental frequency mode-locking state could be sustained up to the maximum available pump power (1.1 W). This mode-locking state has a pulse frequency of 21.7 MHz, rightly consistent with the total cavity length of 4.71 m. At the maximum pump level, the average output power was 268 mW, giving a single pulse energy of 12.37 nJ for the fundamental frequency mode-locking.

After accomplishing the fundamental frequency mode-locking, we carefully tuned both the pump power and the SESAM position to achieve higher-order harmonic mode-locking. Here, harmonic mode-locking transition was obtained through changing the intracavity gain, and different light intensity leaded to different pulse dynamics [64]. Harmonic mode-locking from the first to the fourth order was consecutively observed, as presented in Figure 12 [63]. Compared to the fundamental mode-locking, the pump power had to be increased to over 708 mW to achieve the higher-order harmonic mode-locking. This clearly indicates that more gain is required for sustaining high-order harmonic mode-locking than the fundamental frequency mode-locking.

RF spectrum of different harmonic orders at their maximum output powers is shown in Figure 13. No clear supermode noise was observed for the fundamental and the fourth harmonics, but certain supermode noises were present for the second and the third harmonics. The SNR for the fourth harmonic mode-locking state is ~38 dB.

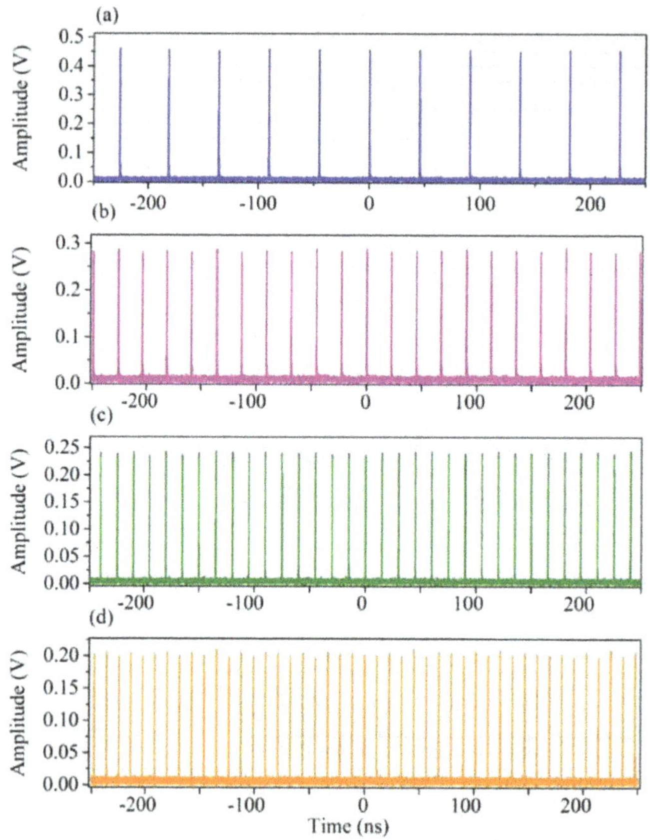

Figure 12. DS pulse trains for (a) fundamental HML at 21.7 MHz, (b) second-order HML at 43.4 MHz, (c) third-order HML at 65.1 MHz, and (d) fourth-order HML at 86.8 MHz [63].

Here, the harmonic transfer was achieved through changing the light intensity, which is different from those based on polarization variation [61, 62]. By tuning the SESAM, we change the light spot size incident on the SESAM and therefore change the light intensity and thus nonlinear phase shift. In addition, the cavity loss is also altered through tuning the SESAM. At the appropriate pump level, careful balancing nonlinearity and dispersion, and gain and loss, finally leads to different harmonic mode-locking states.

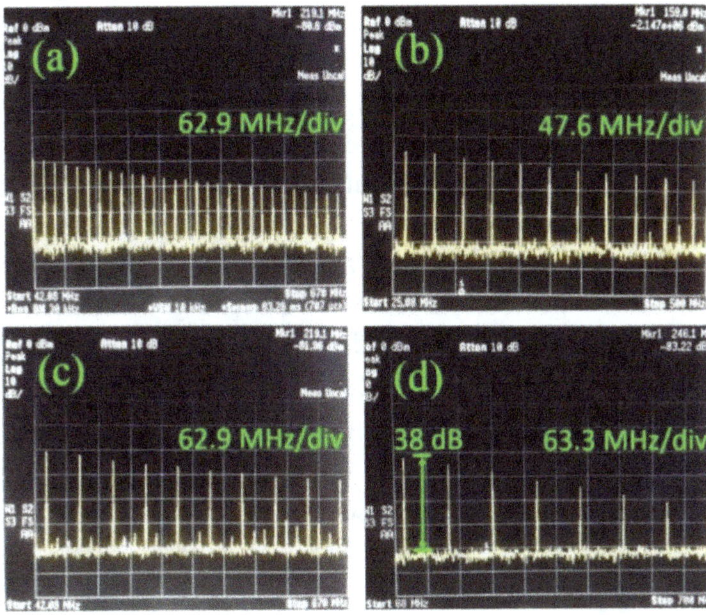

Figure 13. RF spectrum of the fiber laser for the (a) first-, (b) second-, (c) third-, and (d) fourth-order harmonics [63].

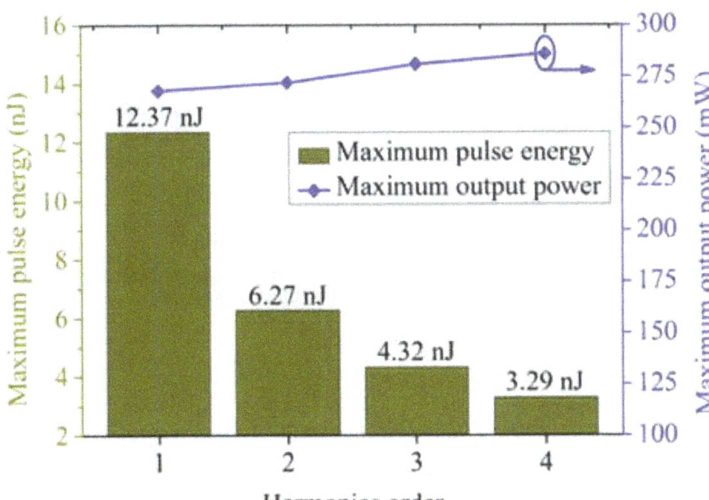

Figure 14. Maximum single pulse energy and average output power of the 2 µm harmonic mode-locked fiber laser at several harmonic orders [63].

In each harmonic mode-locking state, we increased the pump power to the maximum available level, measured the maximum output power, and calculated the corresponding maximum pulse energy, and the results are shown in Figure 14. The highest single pulse energy is 12.37 nJ, achieved with the first-order harmonics. With increasing harmonic order, the pulse energy decreases significantly due to that more pulses (every round trip) need to share the laser power. For the fourth-order harmonics, the maximum single pulse energy is 3.29 nJ. We have also observed the sixth- and eighth-order harmonics mode-locking, but they were not stable. This is probably because that the available pump power (thus gain) is not high enough to balance the loss. Therefore, if higher pump power is provided, higher-order 2 μm HML DSs are expected.

Figure 15(a) shows the laser spectrum of the fundamental pulsing and the successive three high-order harmonics measured at their maximum output levels. All these spectra are similar and have a near-triangle shape and center wavelength of 1929 nm. With increasing harmonic order, spectral width tends to narrow a little bit, which is consistent with the theoretical predictions [65] and experimental results [62, 66]. For the first harmonics, the FWHM width is 3.26 nm, while for the fourth harmonics, the FWHM width is decreased to ~2.5 nm. To get more insight of the pulsing characteristics, we measured autocorrelation (AC) traces of both the fundamental mode-locked DSs and high-order harmonics with an autocorrelator. We found that the AC traces of all these harmonics have similar shape and width and the AC of the fundamental mode-locking pulse is present in Figure 15(b). As shown, the AC trace has a FWHM width of 46.37 ps, corresponding to pulse width of ~30 ps when a $sech^2$ pulse shape is assumed. For the first-order harmonics (fundamental mode-locking), the time-bandwidth product (TBP) of the pulse is calculated to be close to 8, indicating that the pulse was moderately chirped.

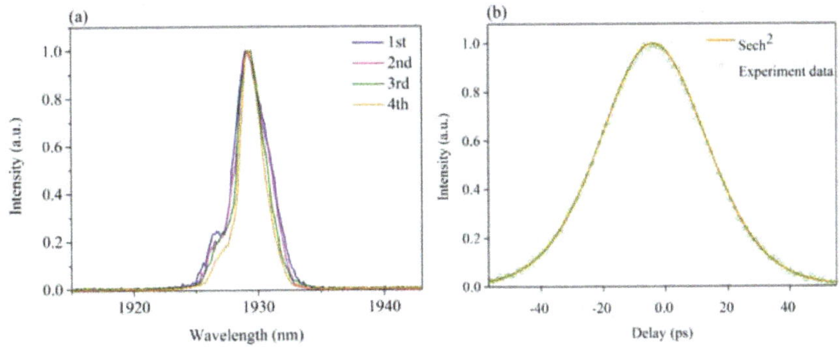

Figure 15. (a) Spectra of the first- to fourth-order harmonic mode-locking solitons and (b) autocorrelation trace of the first-order harmonics [63].

Under every stable mode-locking state, carefully tuning the SESAM can lead to another novel multi-soliton state, the soliton molecule mode-locking. Figure 16 shows two typical kinds of soliton molecules, doublet and the triplet soliton molecules. Soliton molecule is formed through soliton splitting and strong interaction between separated solitons. The total energy of a soliton molecule entity is proportional to the number of single-soliton constituents [67], but the single-soliton's energy is actually decreased. Under, respectively, maximum pump powers (624 and 660 mW), the output powers of the doublet and the triplet soliton molecules are 72 and 95 mW, respectively, corresponding to soliton molecule energies of 3.32 and 4.38 nJ. While their single-soliton energies are 1.66 and 1.46 nJ, respectively.

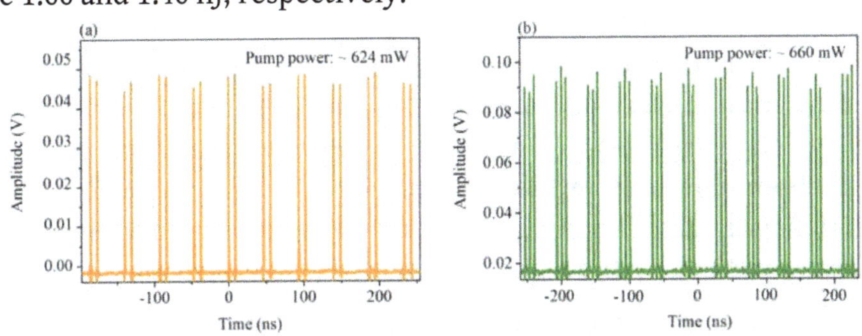

Figure 16. Experimentally measured doublet (a) and triplet (b) soliton molecule pulse trains [63].

5. DISSIPATIVE SOLITON 2 MM FIBER LASERS MODE-LOCKED WITH 2D MATERIALS

Although 2 μm Tm^{3+}-doped fiber lasers (TDFLs) have valuable applications in sensing, medical surgery, industrial machining, and scientific experiments [68, 69], applications require high-peak-power and/or high-energy laser pulses, which are generally produced by using Q-switching [45, 70] or mode-locking [43, 44] methods. Compared with Q-switching, mode-locking can provide much narrower pulse duration and higher peak power.

Passive mode-locking is usually the preferred choice to get short pulses from 2 μm TDFLs, especially with the maturely developed semiconductor SAs as modulators [47]. However, semiconductor SA has some drawbacks such as complex design and growth procedure [71] and narrow working wavelength range. Recently, graphene (a monolayer of two-dimensional (2D) carbon atoms in a honeycomb structure) has attracted great attention for mode-locking of 2 μm TDFLs [72, 73] due to its advantages of large absorption [74], wide operation spectral range [75], and ultrafast recovery time [76]. Another kind of 2D material MoS_2 has also been extensively explored to mode-lock fiber lasers [57, 77, 78, 79, 80]. Although monolayer MoS_2 is a direct band semiconductor (the bandgap determines the energy of photons to be absorbed), studies have proven that layered MoS_2, through introducing stoichiometric defects (non-ideal atomic ratio), also possesses wideband absorption and saturable absorption features [81]. Thereafter, extensive researches have been dedicated to exploring of mode-locking operation and related characteristics of fiber lasers in the 1 μm [57, 77] and 1.5 μm [78, 79, 80] wavelength regions. Owing to large anomalous dispersion of the gain fiber and lower absorption of layered MoS_2 at 2 μm, mode-locking operation with this kind of 2D material and corresponding behavior in the 2 μm region still need further verification.

Here, through combining the CGFML and multilayer MoS_2, we show that mode-locking capability of layered MoS_2 sheets can be definitely extended to the 2 μm wavelength region. With a linear cavity incorporated with the multilayer MoS_2 modulator, fundamental mode-locking in the DS regime for 2 μm Tm^{3+} fiber lasers is achieved. At the same time, through elongating the total fiber length, thus decreasing the mode-locking repetition rate, the pulse energy can be scaled to over 15 nJ.

The multilayer MoS_2 was synthesized with the liquid-phase exfoliation method (LPE) [6], and the MoS_2 nanosheet was transferred onto a gold mirror acting as SA. Raman spectrum of the MoS_2 on the mirror was detected with a spectrometer, and the results are shown in Figure 17(a). The spectral position of the $E_{2g}^{\ 1}$ and A_{1g} modes (~383 cm^{-1} for $E_{2g}^{\ 1}$ and ~408 cm - 1 for A_{1g}) shows that the MoS_2 sample has a thickness of approximately four layers [82]. With a self-constructed 1940 nm ~800 ps fiber laser as the probe source, the reflection method was used to measure the saturable absorption of the sample, and the transmittance of the multilayer MoS_2 on the gold mirror is shown in Figure 17(b) [6]. The nonlinear optical parameters were obtained by using a simple saturable absorption model of [57] here, T(I) is the transmission, α_0 is the modulation depth, I is the input intensity, I_{sat} is the saturation intensity, and α_{ns} is the non-saturable absorbance.

$$T(I) = 1 - \alpha_0 \times \exp(-I/I_{sat}) - \alpha_{ns} \tag{4}$$

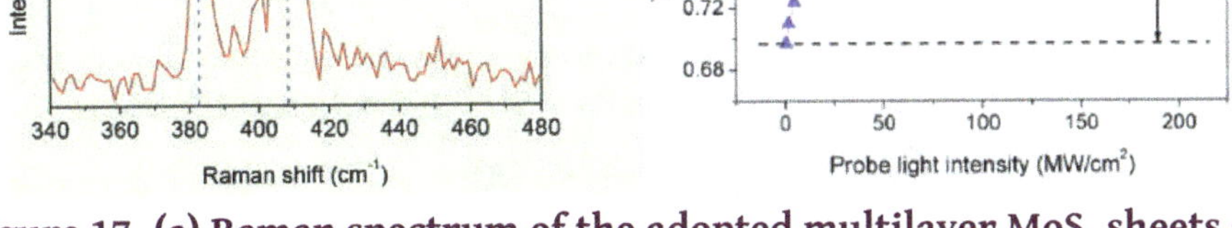

Figure 17. (a) Raman spectrum of the adopted multilayer MoS_2 sheets and (b) nonlinear absorption of the multilayer MoS2 sheets coated on a gold mirror [6].

The measured modulation depth α_0, non-saturable loss α_{ns}, and saturation intensity I_{sat} were 13.6%, 16.7%, and 23.1 MW cm^{-2}, respectively. The modulation depth is comparable to that measured in the 1 μm region [57, 77] but larger than that in the 1.5 μm region [79, 80]. This large modulation depth of the MoS_2 SA at the 2 μm wavelength region is efficient for suppressing wave breaking in mode-locking operation [83].

The schematic diagram of the experimental setup for the MoS_2 mode-locked TDFL is shown in Figure 18. A 1550-nm-CW Er/Yb-codoped fiber laser with maximum output of ~1 W was used as the pump source, and a WDM coupler was used to launch the pump light (with an efficiency of ~95%). The Tm^{3+}-doped silica gain fiber (5/125 μm, 0.24 NA) has core absorption of ~350 dBm^{-1} at ~1550 nm, and 12 cm length of gain fiber was adopted. The dispersion of the gain fiber at 1.9 μm is -12 $ps^2\ km^{-1}$. A 4 m length of SMF-28 fiber was spliced at the output end. To provide normal dispersion, 4.6 m dispersion-compensating fiber (DCF) (2.2 μm, 0.35 NA core) was spliced to the gain fiber. The dispersions of the DCF fiber and the SMF-28 fiber at 1.9 μm are 93 and - 67 $ps^2\ km^{-1}$, respectively [52]. The total net cavity dispersion is ~0.05 ps^2. The DCF fiber was butt coupled to the MoS_2 sheet, which was transferred onto a high-reflection gold mirror. High reflection of the gold mirror and the ~3.5% Fresnel reflection of the perpendicularly cleaved output fiber facet completed the laser cavity.

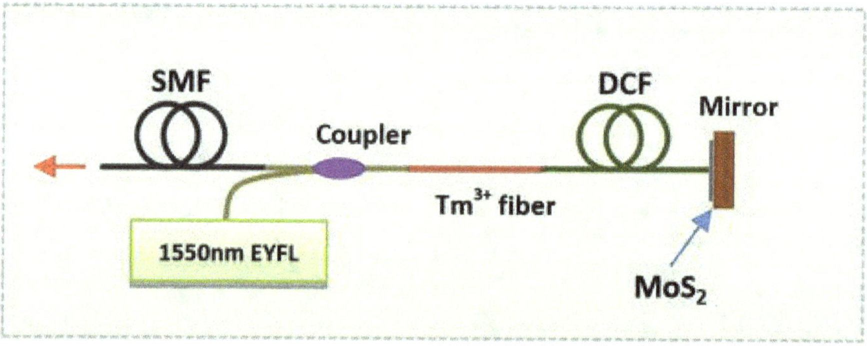

Figure 18. Experimental setup of the mode-locked Tm^{3+} fiber laser. EYFL, erbium/ytterbium-codoped fiber laser; SMF, single-mode fiber; DCF, dispersion-compensating fiber [6].

Under pumping, the 2 μm laser first went to CW operation when pump power was over 430 mW. When the pump power was increased to over 630 mW, the laser came to the stable Q-switching regime. Then, further raising the pump power to over 700 mW and careful adjusting the MoS_2 position, stable mode-locking operation of the TDFL occurred, which could be sustained up to the available maximum pump power. The output power is linearly dependent on the pump power, and the maximum output power is 150 mW, as shown in Figure 19 [6]. The slope efficiency is 43.6% with respect to pump power. The laser spectrum of the mode-locked TDFL is centered at ~1905 nm with a FWHM bandwidth of 17.3 nm. This spectral width is much larger than that of the 1 and 1.5 μm counterparts [57, 77, 78, 79], showing potential much narrower transform-limited pulse duration of this mode-locked TDFL.

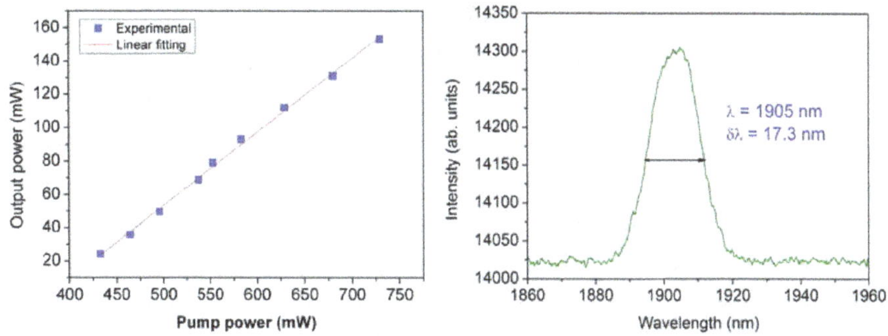

Figure 19. Output (left) and spectrum (right) of the mode-locked Tm^{3+} fiber laser. Square dots are measured data and the solid line is linear fitting [6].

The laser pulse trains obtained at the maximum output level are shown in Figure 20(a) [6]. The 103.4 ns period time corresponds well to the cavity round trip time (the total fiber length is ~10 m), showing that the mode-locking operates at the fundamental frequency of 9.67 MHz. The intensity stability between different pulses is >95%. Considering the 150 mW output power, single pulse energy reaches 15.5 nJ. This is the highest pulse energy ever achieved in mode-locked 2 μm fiber lasers with MoS_2 modulators, and this also demonstrates that 2D material MoS_2 has a great potential in high-power photoelectronics and integrated photonics.

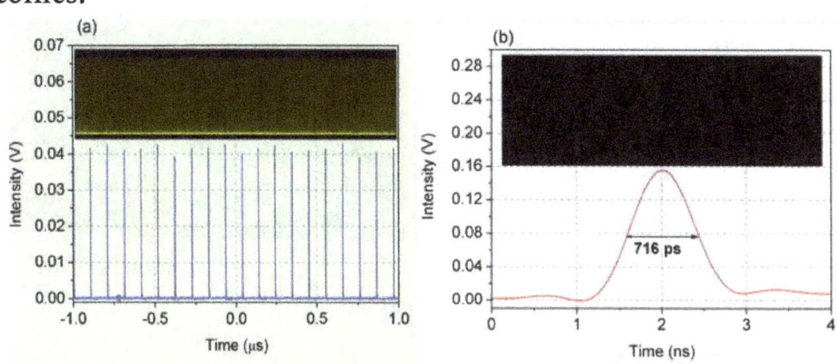

Figure 20. Laser pulse train (a) and single pulse (b) of the MoS_2 mode-locked Tm^{3+} fiber laser measured at the maximum output level. Insets show the oscilloscope traces [6].

Figure 20(b) displays the single pulse at the maximum power level, which has a Gaussian shape and a FWHM width of 716 ps. This 2 μm DS pulse width is comparable to the 1 μm counterparts [57, 77]. Combined with its spectral width, the 2 μm DS pulse has a time-bandwidth product of ~1200, indicating that the mode-locked laser pulse is highly chirped. Chirping pulse also somehow contributes to high pulse energy. In fact, 2D MoS_2 (monolayer or few-layer) has ultrafast recovery times of tens of femtoseconds [84] and ~100 ps [85, 86], corresponding, respectively, to the intraband transition and interband transition of excited free carriers. Recently, mode-locking with similar multilayer MoS_2 SAs has achieved femtosecond time-scale pulse durations [79, 80, 87]. Based on the pulse spectral width (17.3 nm) in our experiment, a Fourier transform-limited pulse width of ~247 fs is expected provided that the entire pulse chirp can be compensated.

We also measured the RF spectrum of this MoS_2 mode-locked fiber laser, and the RF spectrum (with resolution of 0.1 MHz) is shown in Figure 21 (left panel). The fundamental pulsing frequency is 9.67 MHz, which is correspondent to the total cavity length. Over the 100 MHz range, no other supermode oscillations are present. The right panel of the figure shows the ninth-order harmonics in a smaller frequency window (20 MHz), and the signal-to-noise ratio is also >40 dB, showing that the MoS_2 mode-locked fiber laser is comparatively stable.

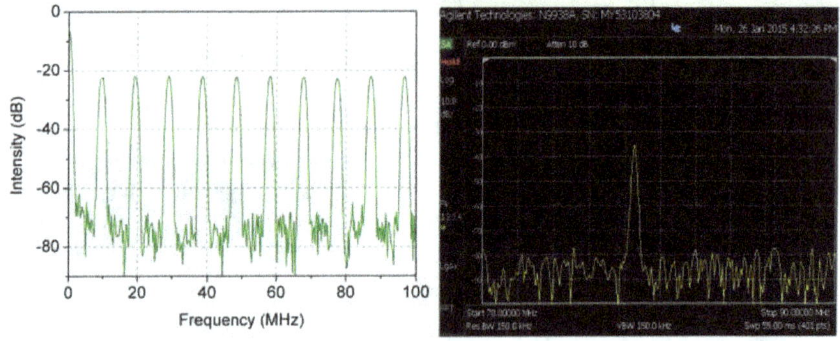

Figure 21. Radiofrequency spectral profile of the mode-locked Tm^{3+} fiber laser [6].

6. CONCISE SUMMARY

Ultrafast 2 μm fiber lasers have gained significant global interest in recent years and have been widely used in various fields such as LIDAR, surgical procedures, molecule spectroscopy, optical sensing, medical treatment, material processing, and nonlinear microscopy [1, 2, 3, 4, 5, 6, 7, 8]. Ultrashort pulses in applications typically necessitate a high pulse energy, which holds significant importance for scientific and industrial purposes. Laser scientists persistently strive to achieve high-energy brief pulses at diverse wavelengths.

Fiber lasers (FLs) are superior to traditional solid-state lasers (SSLs) for generating ultrashort laser pulses because they are more compact, resilient, and produce high-quality laser beams. Traditionally, the soliton mode-locking technique is used to generate brief pulses in fiber networks. Several passive mode-locking approaches can be utilized, such as nonlinear polarization rotation (NPR) [9, 10], nonlinear loop mirror [11, 12], and saturable absorber method [13, 14]. Nevertheless, the pulse energy of conventional solitons, which rely on the equilibrium between dispersion and nonlinearity and have anomalous net cavity dispersion, is typically constrained to less than 1 nJ due to the soliton area theorem [15, 16] or the pulse peak power clamping effect [17, 18]. Thus, fiber lasers continue to generate significantly lower pulse energy compared to solid-state lasers [19].

Various techniques have been proposed and explored to enhance the pulse energy of fiber lasers. These techniques include dispersion-managed soliton, all normal dispersion mode-locking, self-similar soliton, and dissipative soliton. Mechanisms such as dispersion-managed soliton, all normal dispersion mode-locking, self-similar soliton, and dissipative soliton have been found to be particularly effective in improving pulse energy. These mechanisms have been extensively studied and their effectiveness has been demonstrated in multiple studies [20, 21, 22, 23, 24, 25, 26, 27, 28, 29, 30, 31, 32, 33, 34, 35, 36, 37, 38, 39]. DS mode-locked fiber lasers have achieved pulse energy that is 10 to 100 times more than traditional soliton mode-locking lasers by utilizing the balance of nonlinearity, dispersion, gain, and loss [31, 32]. Despite the fact that the pulse energy of 1 to 1.5 μm fiber lasers has surpassed 10 nJ [40, 41, 42] and even reached over 20 nJ [33, 34, 35], the pulse energy of 2 μm DS fiber lasers remains relatively low. The reason for this is that the gain fibers (GFs) now available in the 2 μm range exhibit significant anomalous dispersion. As a result, standard soliton mode-locking operation of 2 μm fiber lasers is achieved [43, 44, 45, 46, 47, 48]. Thus, the pulse energy is still determined by the soliton area theorem and restricted by peak power [15, 17].

The utilization of DS mode-locking has become prevalent as an effective technique for enhancing the pulse energy of 2 μm fiber lasers. In order to achieve DS mode-locking, it is necessary to shift the dispersion of the entire cavity into the normal dispersion zone. In order to achieve this objective, several techniques have been suggested, such as the inclusion of a chirped fiber Bragg grating in the cavity to introduce normal dispersion [49], or the integration of specifically engineered dispersion-compensating fibers (DCFs) into the cavity [50, 51, 52]. Nevertheless, these techniques merely enhance the pulse energy to approximately 1 nJ, leaving the vast potential of the DS mode-locking mechanism still unexplored.

In this study, we will introduce a novel model to examine the pulsing dynamics within a 2 μm DS mode-locked fiber laser. We will demonstrate that, unlike their 1 to 1.5 μm counterparts, the pulse energy of 2 μm DS fiber lasers is primarily constrained by the nonlinear phase shift resulting from the gain fiber. Consequently, we suggest that the anomalous dispersive gain fiber should be minimized in length to effectively separate the effects of gain, dispersion, and nonlinearity. The term we use to refer to it is condensed-gain fiber mode-locking (CGFML). Numerical simulations demonstrate that it is easily achievable to have over 10 nJ of dissipative solitons (DSs) at a wavelength of 2 μm by minimizing excessive phase buildup. Subsequently, we conducted experimental manipulation of a 2 μm fiber laser with CGFML. As a result, we achieved a 4.9 nJ DS with a dechirped pulse duration of 579 fs. By further enhancing the cavity, the pulse energy of this mode-locked fiber laser can be increased to more than 12 nJ, which is similar to that of traditional solid-state mode-locked lasers. The study focuses on analyzing the pulsing dynamics of the CGFML fiber laser under high pumping levels, leading to the successful achievement of high-energy harmonic mode-locking. Here, we demonstrate the successful implementation of high-energy mode-locking in a 2 μm fiber laser using newly discovered 2D materials (MoS_2). Additionally, we have achieved a pulse energy of over 15 nJ. The research findings suggest that the CGFML offers a novel approach for the advancement of high-energy fiber lasers with a short pulse duration in the 2 μm wavelength range.

AT A GLANCE

Mid-infrared (mid-IR) lasers have garnered significant global attention in recent years. The 2 μm Tm3+-doped fiber laser (TDFL) has been significant in the advancement of mid-IR laser sources due to its unique emission wavelength that falls between the near-IR and mid-IR range. The potential uses of this technology are many and include sensing, medical surgery, ranging, telecommunications, and as pump sources for the development of 3–5 μm laser systems. Despite achieving a continuous-wave (CW) output power of over 1000 W, the performance of high-pulse-energy ultrafast 2 μm TDFLs is still constrained by nonlinear optical phenomena. In the context of classical soliton mode-locking, the pulse energy is constrained by the soliton area theorem (also known as the energy quantization principle), which sets an upper limit. One potentially effective technique for enhancing the pulse energy of 2 μm fiber lasers is through the use of dissipative soliton (DS) mode-locking. This chapter provides an overview of the latest advancements in high-energy ultrafast DS 2 μm TDFLs generated in our research. It also examines the possibilities and future prospects of this field. Through the implementation of a novel approach called condensed-gain fiber mode-locking, we demonstrate the ability to consistently increase the soliton pulse energy of 2 μm TDFLs to more than 10 nJ. Additionally, this technique allows for the observation of diverse soliton dynamics such as harmonic mode-locking and soliton molecules. Additionally, the process of DS mode-locking of TDFLs using one of the materials that resembles a two-dimensional structure, such as MoS_2, is being examined.

REFERENCES

1. Jackson SD. Towards high-power mid-infrared emission from a fibre laser. Nature Photonics. 2012;6:423-431

2. Geng J, Jiang S. The 2 μm market heats up. Optics and Photonics News. 2014;25:36-41

3. Tang Y, Yu X, Li X, Yan Z, Wang QJ. High-power thulium fiber laser Q switched with single-layer graphene. Optics Letters. 2014;39:614-617

4. Wang Y et al. High power tandem-pumped thulium-doped fiber laser. Optics Express. 2015;23:2991-2998
5. Ferman M. E and Hartl I. Nature Photonics. 2013;7:868
6. Tian Z, Wu K, Kong LC, Yang N, Wang Y, Chen R, Hu W, Xu JQ, Tang YL. Lasers Physics Letters. 2015;12:065104
7. Kang K, Liu MY, Gao XJ, Li N, Yin SY, Qin GS, Qin WP. Lasers Physics Letters. 2015;12:045106
8. Xu C, Wise FW. Nature Photonics. 2013;7:875
9. Huang Y, Xu X. Lasers. Physics Letters. 2014;11:125101
10. Matsas VJ, Richardson DJ, Newson TP, Payne DN. Optics Letters. 1993;18:358
11. Honzatko P, Baravets Y, Todorov F. Lasers Physics Letters. 2013;10:075103
12. Xu B, Martinez A, Set SY, Goh CS, Yamashita S. Lasers Physics Letters. 2014;11:025101
13. Gomes LA, Orsila L, Jouhti T, Okhotnikov OG, IEEE J. Selected Topics in Quantum Electronics. 2004;10:129
14. Choudhary A, Lagatsky AA, Zhang ZY, Zhou KJ, Wang Q, Hogg RA, Pradeesh K, Rafailov EU, Sibbett W, Brown CAT, Shepherd DP. Lasers Physics Letters. 2013;10:105803
15. Hasegawa A et al. Transmission of stationary nonlinear optical pulses in dispersion dielectric fibers. I. Anomalous dispersion. Applied Physics Letters. 1973;23:142-144
16. Kelly S. Electronics Letters. 1992;28:806
17. Zakharov V, Shabat A. Soviet Physics JETP. 1973;37:823
18. Tang D Y, Zhao L M, Zhao B and Liu A Q, Physical Review A. 2005;72:043816
19. Wise FW, Chong A, Renninger WH. High-energy femtosecond fiber lasers based on pulse propagation at normal dispersion. Laser & Photonics Review. 2008;2:58-73

20. Tamura K, Ippen EP, Haus HA, Nelson LE. 77-fs pulse generation from a stretched-pulse mode-locked all-fiber ring laser. Optics Letters. 1993;18:1080-1082

21. Tamura K, Ippen EP, Haus HA. Pulse dynamics in stretched-pulse fiber lasers. Applied Physics Letters. 1995;67:158-160

22. Lim H, Ilday FO, Wise FW. Femtosecond ytterbium fiber laser with photonic crystal fiber for dispersion control. Optics Express. 2002;10:1497-1502

23. Ilday FO, Wise FW. Nonlinearity management: A route to high-energy soliton fiber lasers. Journal of Optical Society America B. 2002;19:470-476

24. Tang DY, Zhao LM, Xie GQ, Qian LJ. Coexistence and competition between different soliton-shaping mechanisms in a laser. Physical Review A. 2007;75:063810

25. Ilday FO, Buckley JR, Lim H, Wise FW, Clark WG. Generation of 50-fs, 5-nJ pulses at 1.03 μm from a wave-breaking free fiber laser. Optics Letters. 2003;28:1365-1367

26. Buckley JR, Wise FW, Ilday FO, Sosnowski T. Femtosecond fiber lasers with pulse energies above 10 nJ. Optics Letters. 2005;30:1888-1890

27. Ilday FO, Buckley JR, Clark WG, Wise FW. Self-similar evolution of parabolic pulses in a laser. Physical Review Letters. 2004;92:213902

28. Renninger WH, Chong A, Wise FW. Self-similar pulse evolution in an all-normal-dispersion laser. Physical Review A. 2010;021805(R):82

29. Liu H, Liu Z, Lamb ES, Wise F. Self-similar erbium-doped fiber laser with large normal dispersion. Optics Letters. 2014;39:1019-1021

30. Oktem B, Ulgudur C, Ilday FO. Soliton-similariton fibre laser. Nature Photonics. 2010;4:307-311

31. Chong A, Renninger WH, Wise FW. All-normal-dispersion femtosecond fiber laser. Optics Express. 2006;14:10095-10100

32. Chong A, Renninger WH, Wise FW. Properties of normal-dispersion femtosecond fiber lasers. Journal of the Optical Society of America B: Optical Physics. 2008;25:140-148

33. Chong A et al. All-normal-dispersion femtosecond fiber laser with pulse energy above 20 nJ. Optics Letters. 2007;32:2408-2410
34. Kieu K, Renninger WH, Chong A, Wise FW. Sub-100 fs pulses at wattlevel powers from a dissipative-soliton fiber laser. Optics Letters. 2009;34:593-595
35. Chichkov NB et al. High-power dissipative solitons from an all-normal dispersion erbium fiber oscillator. Optics Letters. 2010;35:2807-2809
36. Bale B, Boscolo S, Turitsyn S. Dissipative dispersion-managed solitons in mode-locked lasers. Optics Letters. 2009;(21):3286-3288
37. Renninger WH, Wise FW. IEEE Journal of Selected Topics in Quantum Electronics. 2015;21:101
38. Chichkov NB, Hapke C, Neumann J, Kracht D, Wandt D, Morgner U. Optics Express. 2012;(4):3844
39. Zhao LM, Tang DY, Wu J. Optics Letters. 2006;31:1788
40. Baumgartl M, Ortaç B, Lecaplain C, Hideur A, Limpert J, Tünnermann A. Optics Letters. 2010;35:2311
41. Mao D, Liu XM, Wang RL, Hu XH, Lu H. Laser Physics Letters. 2011;8:134
42. Choi SY, Jeong H, Hong BH, Rotermund F, Yeom D. Laser Physics Letters. 2014;11:015101
43. Nelson LE, Ippen EP, Haus HA. Broadly tunable sub-500 fs pulses from an additive-pulse mode-locked thulium-doped fiber ring laser. Applied Physics Letters. 1995;67:19-21
44. Sharp RC, Spock DE, Pan N, Elliot J. 190-fs passively mode-locked thulium fiber laser with a low threshold. Optics Letters. 1996;21:881-883
45. Solodyankin MA et al. Mode-locked 1.93 μm thulium fiber laser with a carbon nanotube absorber. Optics Letters. 2008;33:1336-1338
46. Wang Q, Geng J, Luo T, Jiang S. Mode-locked 2 μm laser with highly thulium-doped silicate fiber. Optics Letters. 2009;34:3616-3618

47. Wang Q, Geng J, Jiang Z, Luo T, Jiang S. Mode-locked tm–Ho-codoped fiber laser at 2.06 μm. IEEE Photonics Technology Letters. 2011;23:682-684

48. Wang Q, Chen T, Zhang B, Heberle AP, Chen KP. All-fiber passively mode-locked thulium-doped fiber ring oscillator operated at solitary and noiselike modes. Optics Letters. 2011;36:3750-3752

49. Gumenyuk R, Vartiainen I, Tuovinen H, Okhotnikov OG. Dissipative dispersion-managed soliton 2 μm thulium/holmium fiber laser. Optics Letters. 2011;36:609-611

50. Agrawal GP. Nonlinear Fiber Optics. Academic Press; 2007

51. Haxsen F, Wandt D, Morgner U, Neumann J, Kracht D. Monotonically chirped pulse evolution in an ultrashort pulse thulium-doped fiber laser. Optics Letters. 2012;37:1014-1016

52. Wang QQ et al. All-fiber ultrafast thulium-doped fiber ring laser with dissipative soliton and noise-like output in normal dispersion by single-wall carbon nanotubes. Applied Physics Letters. 2013;103:011103

53. Huang C, Wang C, Shang W, Tang Y, Xu J. Scientific Reports. 2015;5:13680

54. Renninger WH, Wise FW. Pulse shaping and evolution in normal-dispersion mode-locked fiber lasers. IEEE Journal of Selected Topics in Quantum Electronics. 2012;18:389-398

55. Cabasse A. 130 mW average power, 4.6 nJ pulse energy, 10.2 ps pulse duration from an Er3+ fiber oscillator passively mode locked by a resonant saturable absorber mirror. Optics Letters. 2011;36:2620-2622

56. Yang N, Huang CY, Tang Y, Xu J. Lasers Physics Letters. 2015;055101:12

57. Du J, Wang Q, Jiang G, Xu C, Zhao C, Xiang Y, Chen Y, Wen S, Zhang H. Scientific Reports. 2014;4:6346

58. Chu SW, Chen SY, Tsai TH, Liu TM, Lin CY, Tsai HJ, Sun CK. Optics Express. 2003;11:3093

59. Yilmaz T, Depriest CM, Turpin T, Abeles JH, Delfyett PJ. IEEE Photonics Technology Letters. 2002;14:1608

60. Richardson DJ, Nilsson J, Clarkson WA. Journal of the Optical Society of America B: Optical Physics. 2010;27:B63

61. Yan PG, Lin RY, Ruan SC, Liu AJ, Chen H. Optics Express. 2015;23:154

62. Peng JS, Zhan L, Luo SY, Shen QS. Journal of Lightwave Technology. 2013;31:3009

63. Yang N, Tang YL. Xu J, Lasers. Physics Letters. 2015;12:085102

64. Kutz JN, Sandtede B. Optics Express. 2008;16:636

65. Haboucha A, Komarov A, Leblond H, Sanchez F, Martel G. Optical Fiber Technology. 2008;14:262

66. Chen H, Chen SP, Jiang ZF, Hou J. Optics Express. 2015;23:1308

67. Akhmediev N, Grelu P. Nature Photonics. 2012;6:84

68. Zhang Y, Tian Y, Wang W, Yao B. Tunable narrow linewidth Tm^{3+}-doped silica fiber laser with an intracavity taper. Laser Physics Letters. 2010;7:225

69. Wang F, Shen DY, Fan DY, Lu QS. High power widely tunable Tm^{3+}: Fiber laser with spectral linewidth of 10 pm. Laser Physics Letters. 2010;7:450

70. Eichhorn M, Jackson SD. High-pulse-energy actively Q-switched Tm^{3+}-doped silica 2 μm fiber laser pumped at 792 nm. Optics Letters. 2007;32:2780

71. Saraceno CJ, Schriber C, Mangold M, Hoffmann M, Heckl OH, Baer CRE, Golling M, Südmeyer T, Keller U. SESAMs for high-power oscillators: Design guidelines and damage thresholds. IEEE Journal of Selected Topics in Quantum Electronics. 2012;18:29

72. Sobon G, Sotor J, Pasternak I, Krajewska A, Strupinski W, Abramski KM. Thulium-doped all-fiber laser mode-locked by CVD-graphene/PMMA saturable absorber. Optics Express. 2013;21:12797

73. Wang Q, Chen T, Zhang B, Li M, Lu Y, Chen KP. All-fiber passively mode-locked thulium-doped fiber ring laser using optically deposited graphene saturable absorbers. Applied Physics Letters. 2013;102:131117

74. Nair RR, Blake P, Grigorenko AN, Novoselov KS, Booth TJ, Stauber T, Peres NMR, Geim AK. Fine structure constant defines visual transparency of grapheme. Science. 2008;320:1308

75. Zhang YZ, Liu T, Li XH, Liang G, Meng B, Hu X, Wang QJ". Broadband high photoresponse from pure monolayer graphene-based photodetector. Nature Communications. 2013;4:1811

76. Bao Q, Zhang H, Wang Y, Ni Z, Yan Y, Shen ZX, Loh KP, Tang DY. Atomic-layer graphene as a saturable absorber for ultrafast pulsed lasers. Advanced Functional Materials. 2009;19:3077

77. Zhang H, Lu SB, Zheng J, Du J, Wen SC, Tang DY, Loh KP. Molybdenum disulfide (MoS2) as a broadband saturable absorber for ultra-fast photonics. Optics Express. 2014;22:7249

78. Xia H, Li H, Lan C, Li C, Zhang X, Zhang S, Liu Y. Ultrafast erbium-doped fiber laser mode-locked by a CVD grown molybdenum disulfide (MoS2) saturable absorber. Optics Express. 2014;22,17341

79. Liu H, Luo A, Wang F, Tang R, Liu M, Luo Z, Xu W, Zhao C, Zhang H. Femtosecond pulse erbium-doped fiber laser by a few-layer MoS2 saturable absorber. Optics Letters. 2014;39:4591

80. Khazaeizhad R, Kassani SH, Jeong H, Yeom D, Oh K. Mode-locking of Er-doped fiber laser using a multilayer MoS2 thin film as a saturable absorber in both anomalous and normal dispersion regimes. Optics Express. 2014;22,23732

81. Wang S, Yu H, Zhang H, Wang A, Zhao M, Chen Y, Mei L, Wang J. Broadband few-layer MoS2 saturable absorbers. Advanced Materials. 2014;26:3538

82. Li H, Zhang Q, Yap CCR, Tay BK, Edwin THT, Olivier A, Baillargeat D. From bulk to monolayer MoS2:Evolution of Raman scattering. Advanced Functional Materials. 2012;22:1385

83. Cabasse A, Martel G, Oudar JL. High power dissipative soliton in an erbium-doped fiber laser modelocked with a high modulation depth saturable absorber mirror. Optics Express. 2009;17:9537

84. Wang K et al. Ultrafast saturable absorption of 2D MoS2 nanosheets. ACS Nano. 2013;7:9260

85. Korn T, Heydrich S, Hirmer M, Schmutzler J, Schuller C. Low-temperature photocarrier dynamics in monolayer MoS2. Applied Physics Letters. 2011;99:102109

86. Wang R, Ruzicka BA, Kumar N, Bellus MZ, Chiu HY, Zhao H. Ultrafast and spatially resolved studies of charge carriers in atomically thin molybdenum disulphide. Physical Review B. 2012;86:045406

87. Wu K, Zhang X, Wang J, Chen J. 463 MHz fundamental mode-locked fiber laser based on few-layer MoS2 saturable absorber. Optics Letters. 2015;40:1374

CHAPTER-7

GENERATION AND DIAGNOSTICS OF HIGH-POWER ATTOSECOND PULSES: A FREQUENCY-SYNTHESIZED APPROACH

1. INTRODUCTION: AN OVERVIEW

The history of humanity has demonstrated that the development of innovative light sources and related technologies would result in significant scientific advancements and have a profound impact on society and civilization. Lasers and X-rays are prime examples of such technologies. A class of coherent light sources known as high-power laser systems is instrumental in the advancement of science and technology. These systems are used in a variety of applications, including inertial nuclear fusion, laboratory astrophysics, laser weapons, and laser 3D printing. Continuous wave (CW), nanosecond (ns), picosecond (ps), and femtosecond (fs) pulsed output are all produced by these lasers. Laser-based methods can now produce single or near-single cycle electromagnetic radiation at frequencies ranging from the terahertz (1 THz = 1012 Hz) to the soft X-ray regions of the spectrum. Attosecond pulses (1 as = 10–18 s) are produced by the spectra of the latter. It is anticipated that these innovative sources will have a diverse array of potential applications. Attosecond sources [1, 2] are among the most thrilling new laser sources that are presently in the process of development. Electrons can be directed within and around atoms by controlled light waves in the near future.

This emerging technology has been referred to as "lightwave electronics" [3]. However, the investigation of condensed matter with attosecond time-resolution continues to be a challenge [4]. Although the potential of ultrafast lasers to facilitate microfabrication and nanostructuring of materials was acknowledged and demonstrated over a decade ago [5], there have been no reports of real-world applications of attosecond pulses to date. This is primarily restricted by the absence of potent attosecond sources.

High-order harmonic generation (HHG) [6] appears to be the most viable approach among those that enable the generation of attosecond pulses. HHG has the potential to generate attosecond pulses of high intensity that extend from the Vacuum Ultraviolet (VUV) or extreme ultraviolet (EUV) to the soft X-ray region [7]. Conversely, Chen et al. [8] and Hsieh et al. [9] demonstrated that the high-order stimulated Raman scattering (HSRS) process can generate a sub-cycle pulse train that is governed by the carrier-envelope-phase (CEP). We have recently demonstrated the generation of attosecond pulses by synthesising the harmonics of the same laser up to the fifth order. These harmonics were produced through second-order nonlinear optical processes, such as second harmonic generation (SHG) or sum frequency generation (SFG). With carrier-envelope-phase (CEP) control, this innovative source is capable of producing sub-single-cycle (~0.37 cycle) pulses with a peak intensity of a single pulse as high as 1014W/cm^2 and a pulse width as brief as 400 attosecond [10]. Figure 1 illustrates the intensity (red traces) and waveform (purple trace) of these ultrashort pulses.

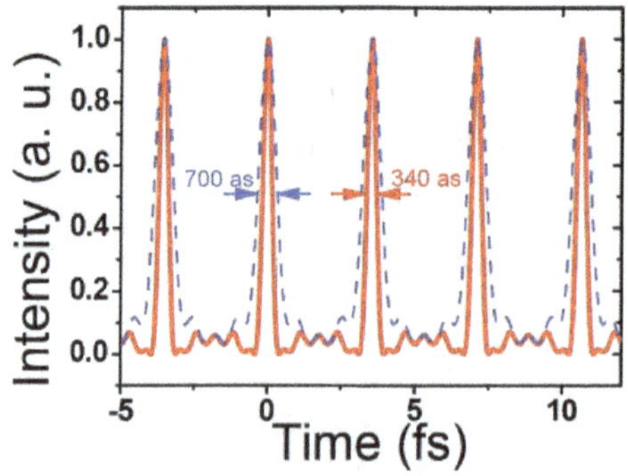

Figure 1. Waveform (blue trace) and intensity (red traces) of sub-femtosecond pulses synthesized by cascaded harmonics of an injection-seed high-power Q-switched laser.

In addition, we demonstrate that the optical fields of the harmonics can be maintained at a constant relative phase for a minimum of thousands of nanosecond pulses. The absolute maximum relative phase fluctuation is 0.04 π rad. This method demonstrates the ability to produce sub-femtosecond (360 attosecond) pulses with carrier-envelope phase (CEP) control. Arbitrary waveforms, such as square and sawtooth waveforms, can be synthesized [11].

This novel light source offers the advantages of simplicity and compactness in comparison to the conventional method of generating attosecond pulses through higher-harmonic generation (HHG) of few-cycle femtosecond pulses. Additionally, the attosecond pulse generated in this manner is sub-single-cycle with full CEP control, and an arbitrary optical waveform can be synthesized. The fundamental principle for the generation of attosecond pulses through the synthesis of cascaded harmonics is presented in Section 2 of the chapter. The prototype system is described. The diagnostics of such broadband sources are nontrivial. Section 3 presents experimental methodologies for relative phase control among the harmonics. This is succeeded by an examination of the linear cross-correlation method's application to the synthesis of arbitrary waveforms and their diagnostics. This chapter concludes with a summary in Section 5. Part 2 of this work (the subsequent chapter) contains applications of this innovative high-power laser system. In this section, we explore coherently controlled harmonic generation [12] and phase-sensitive 2-color ablation of copper and stainless steel using this multi-color laser system.

2. GENERATION OF ATTOSECOND PULSES BY SYNTHESIS OF CASCADED HARMONICS

Fundamentally, an optical pulse train with a repetition rate of ωm can be viewed as the sum of a set of frequency components that form an arithmetic series [13]. The electric field of each component can be written in the following form:

$$E_q(t) = A_q e^{i\phi_q} e^{i\omega_q t}, \tag{1}$$

where $\omega_q = \omega_0 + q\omega_m$, for $q = 0, 1, 2, \ldots$ To shape the pulse envelope, the phase term ϕ_q and amplitude term A_q of each component are controlled.

One can set the phase term ϕ_q and rewrite it as $\phi_q = \phi_0 + q\phi_m$. The synthesized pulse could then be expressed as:

$$E(t) = \sum_q E_q(t) = e^{i(\omega_0 t + \phi_0)} \sum_q A_q e^{iq\omega_m \left(t + \frac{\varphi_m}{\omega_m}\right)} = e^{i(\omega_0 t + \phi_0)} E_c\left(t + \varphi_m / \omega_m\right), \quad (2)$$

where $E_c(t) \equiv \sum_q A_q e^{iq\omega_m t}$ is a typical cosine pulse train and $\omega_0 t + \phi_0$ is the time-varying CEP with frequency of ω_0. In the commensurate case, the CEP is equal to ϕ_0 for all ultrashort pulses belonging to the same attosecond pulse train or within the ns pulse envelope in the HSRS approach since ω_0 equals to zero. As a result, CEP will be randomly changing if ϕ_0 is random from 1 ns pulse to another. For instance, a 802 nm and a 602 nm laser with pulsewidth around ns and repetition rate of 30 Hz (corresponding to q = 3 and 4 of the Raman resonance of molecular hydrogen) were employed to stimulate the Raman sidebands in early work by one of the co-authors [14, 15]. Because the phases of the two driving lasers, denoted as ϕ_3 and ϕ_4, are random and independent of each other in individual ns pulses, both $\phi_m = \phi_4 - \phi_3$ and $\phi_0 = \phi_3 - 3\phi_m = 4\phi_3 - 3\phi_4$ are random in time as well. Although CEP of generated pulse trains with 1 ns pulse every 33 ms is fixed. The CEP of attosecond pulses in different ns pulse envelopes varies randomly as well. This severely limits the application of this type of attosecond light source. Cross-correlation by four-wave-mixing interaction among attosecond pulses within the same ns pulse, which are commensurate. Therefore, correlation behavior could still be observed.

Alternatively, CEP will be fixed if all phase-controlled frequency components of the pulse train are optical harmonics from the same laser, rather than through Raman sideband generation. It is clear that the relative phase among generated higher-order harmonics and the lower ones are fixed. For example, relative phase among ϕ_5, ϕ_2 and ϕ_3 of the fifth, second and third harmonic of the same laser will not be changing, if light of frequency ω_5 is generated from SFG of ω_2 and ω_3.

At this junction, it is instructive to note that Hansch proposed that sub-femtosecond pulse could be synthesized by nonlinear phase locking of lasers nearly a decade ago [16]. Later, his group further demonstrated the feasibility of this approach with three cw phase-locked semiconductor lasers [17]. This approach, however, was not pursued since primarily because of the low power generated.

In the following, we summarize the potential advantages and unique features of attosecond pulse generation through pulse synthesis of harmonics of the same laser in contrast to the earlier Raman sideband approach:

2.1. Higher efficiency

This is expected since all the wavelength conversion processes for generation of the harmonics up to the near UV are from second order nonlinearity, instead of third order nonlinearity in the Raman sideband approach.

2.2. Simplicity and compactness

Only one pump laser is required in the present scheme (see Figure 2(b)) rather than two in the Raman sideband approach.

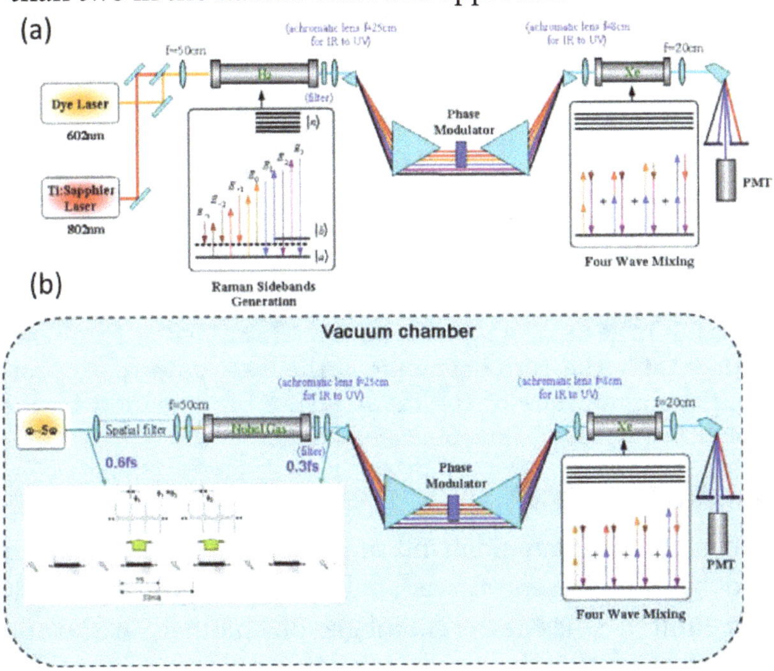

Figure 2. (a) Schematic of the Raman sideband generation approach (b): Schematic of the cascaded harmonics approach A Q-switched Nd:YAG laser and its harmonics up to the fifth order is used as the laser source.

2.3. Role of the fundamental frequency

In frequency-domain description of the pulse train, mode spacing or frequency difference of adjacent modes dictates pulse spacing in time or pulse repetition rate in frequency [13]. Therefore, lower-frequency components of the pulse decide the main structure of the pulse waveform while higher-frequency components provide the fine structure or details of the pulse. To illustrate, Figure 3 shows a square wave synthesized by 10 modes with frequencies from the fundamental to 10th harmonic of the pump laser. Severe distortion is observed if the amplitude of the fundamental is attenuated merely by 10% (see Figure 3(b)). This reveals the significance of the component at the fundamental frequency. Significantly, high-quality square waveform can already be synthesized with five frequency components from the fundamental frequency to fifth harmonic frequency.

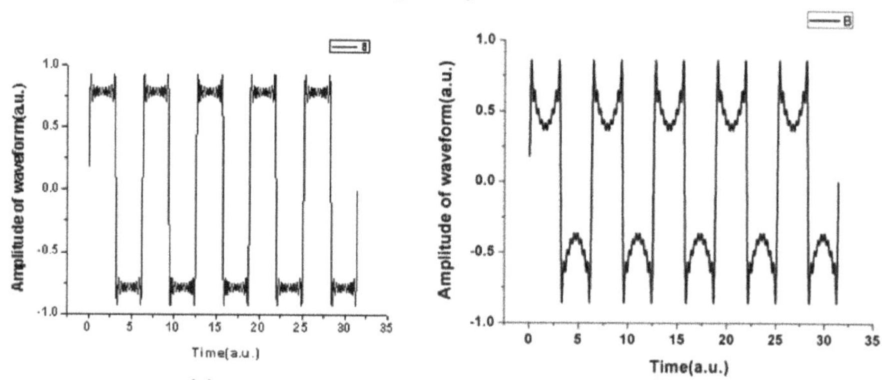

(a) (b)

Figure 3. (a) Synthesis of a square wave with modes at frequency from the fundamental to the 10th harmonic of the laser output. (b) Same as (a) except that the amplitude of the mode at the fundamental frequency is attenuated by 10%.

2.4. Bandwidth

Compared to the Raman sideband approach, shortest pulse duration generated by the present cascaded harmonics synthesis method is inevitably limited since fewer numbers of channels are available in practice. However, this does not severely limit the application of the latter in generating attosecond pulses. A pulse 0.6 fs in duration could be obtained by synthesis of the fundamental wavelength of 1064 nm and its second, third, fourth and fifth harmonics. This can be understood by realizing that bandwidth is independent of the number of channels physically. These five components already span sufficient bandwidths.

2.5. Pulse quality

Arbitrary waveform synthesis is of importance for attosecond science. As an example, we show the synthesis of a Gaussian pulse with various numbers of frequency components or channels. The image-quality-index, which is widely used in image pattern recognition, is used to gauge the quality of the shaped pulse [18]. The quality increases step-wise only when number of channels is equal to 2^n (n is an integer), that is, 2, 4, 8, 16...). This implies that we do not have to put too much effort into generating the sixth and seventh harmonic unless the 8th harmonic (133 nm) can be generated efficiently as well (Figure 4).

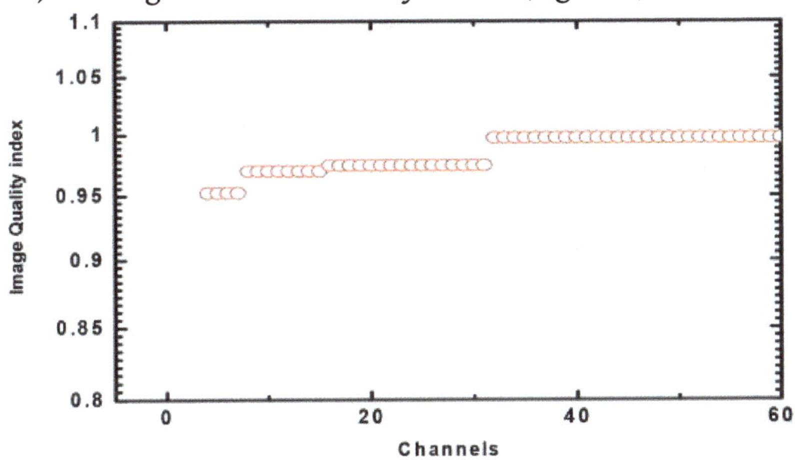

Figure 4. Image quality index of a Gaussian pulse generated with different number of channels.

As a further example, we show the synthesis of a sawtooth waveform with various numbers of channels (see Figure 5(a)). Figure 5(a) also indicates that a quality factor of 92% could be achieved already with 4 channels. The perfect sawtooth wave, those synthesized with 4 channels (from the fundamental to the fourth harmonics) and 32 channels (from the fundamental to the 32nd harmonics) are illustrated in Figure 5(b). The sawtooth waveform synthesized with 4 channels is already recognizable, while that generated with 32 channels is indistinguishable from the mathematical function.

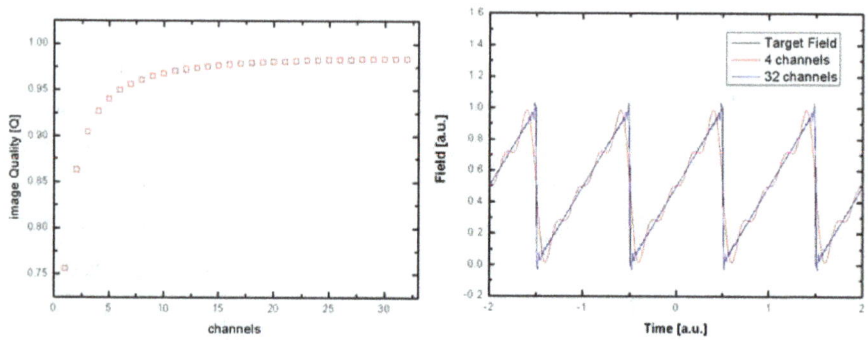

(a) (b)

Figure 5. (a) Synthesis of a sawtooth waveform with various numbers of channels (b) the perfect sawtooth wave, those synthesized with 4 channels (from the fundamental to the fourth harmonics) and 32 channels (from the fundamental to the 32th harmonics).

In our lab, we have constructed a system for the demonstration of attosecond pulse generation by synthesis of cascaded harmonics. This is shown in Figure 6. The fundamental frequency component at ω_1 is from a custom-made injection-seeded Quanta Ray PRO- 290 Q-switched Nd:YAG laser (λ = 1064 nm) operating at 10 Hz. The pulse duration is about 10 ns. The laser frequency bandwidth is narrower than 0.003 cm^{-1}. The nonlinear optical crystals for generating 2nd through the 5th harmonics of the laser fundamental beam are arranged in a cascaded layout. The crystals are KD*P type II for the second harmonic ω_2, KD*P type I for the third harmonic ω_3, BaB_2O_4 (BBO) type I for the fourth harmonic ω_4, and BBO type I for the fifth harmonic ω_5. Thus the five-color output of the laser system covers optical spectra from the near infrared (NIR) or 1064 nm to the ultraviolet (UV), that is, 213 nm. The cascade setup was adopted to ensure that the second-order nonlinear optical process all occurred collinearly. As a result, the fundamental and harmonics overlapped spatially. The pulse energy of each harmonic was 380, 178, 70, 41, and 22 mJ, respectively. The polarizations of the five colors were elliptic, horizontal, vertical, vertical, and horizontal for the fundamental through the fifth harmonic in that order. Eventually, all five colors will be converted into horizontally polarized light (see below).

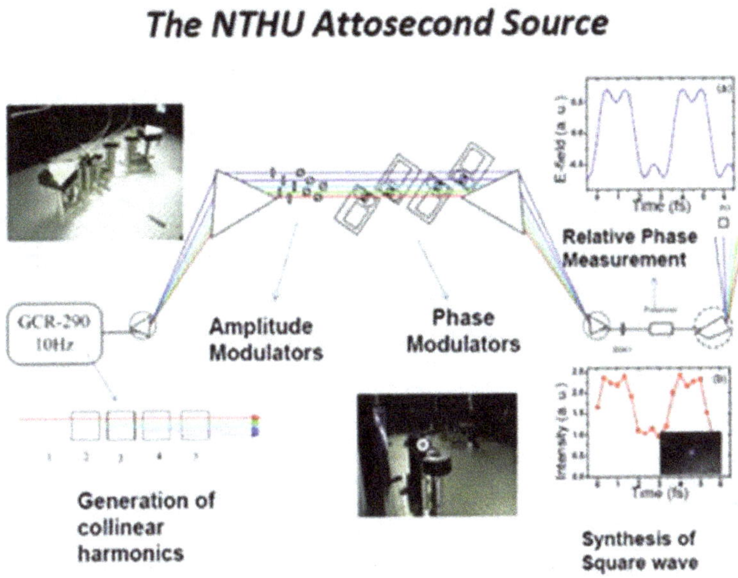

Figure 6. First-generation NTHU Attosecond source based on frequency synthesis of cascaded in-line harmonics of a single-frequency Q-switched Nd:YAG laser. Amplitude and phase modulation of each of the harmonics are provided. Insets (a) and (b) show predicted and experimentally generated square waveforms.

Precision control of the amplitude and the phase of each frequency components are essential. To this end, we first spatially dispersed the five colors by a fused silica prism. The dispersed beams were then recollimated but spatially separated by using another, larger fused silica prism. In the parallel co-propagating region of the five colors, we inserted an amplitude modulator and a phase modulator each. Therefore, it is possible to adjust the amplitude and relative phase of these harmonics separately. Each amplitude modulator was the assembly of a half waveplate and a polarizer. The polarization directions of the harmonic frequencies were all horizontal after passing through the polarizers. We can adjust the pulse energy of each harmonic by rotating the orientation of the half waveplate. To deal with the elliptical polarization of the fundamental frequency of the laser, we used a quarter waveplate to rotate the elliptical polarization back to the linear polarization.

Each phase modulator consisted of a pair of right-angle triangle prisms. Adjusting the relative positions of each prism in the pair independently along the direction of their hypotenuse will change the effective path length traveled by each harmonic. The phase of each harmonic wave will be altered by $\Delta\phi_i = 2\pi\,(n_{prism} - n_{air})l/\lambda_i$, where l is the relative displacement of the two prisms n_{prism}, and n_{air} are refractive indices of the prism and air, respectively. This scheme allows variation of the phase $\Delta\phi_i$ of the i_{th} harmonic, $i = 1–5$, but will not affect the beam alignment. Finally, these five beams of fundamental output and cascaded harmonics of the Nd:YAG laser were recombined and collimated by another prism set at a symmetry position to the first prism set. The whole setup is similar to a 4-f imaging system.

3. RELATIVE PHASE MEASUREMENT

For waveform control and pulse synthesis, we need to determine the relative phase among the harmonics. This was accomplished as follows: First, the fifth harmonic was used as the reference. We then proceed to adjust the relative phase of all other four harmonic frequencies to the reference. Four type I BBO crystals were employed. These were cut at (a) $\theta = 22.9°$ and $\phi = 0°$ for 1064 nm + 1064 nm → 532 nm, (b) $\theta = 31.3°$ and $\phi = 0°$ for 1064 nm + 532 nm → 355 nm, (c) $\theta = 47.7°$ and $\phi = 0°$ for 532 nm + 532 nm → 266 nm, and (d) $\theta = 51.2°$ and $\phi = 0°$ for 1064 nm + 266 nm → 213 nm, respectively. The Nd:YAG laser harmonic frequencies and summed frequencies generated from the BBO crystal were then dispersed and detected by a set of photodiodes. Every harmonic was heterodyned with a signal at the same frequency derived by optically summing two lower harmonics in a particular BBO crystal. Then the resulting interference signal can be used to calibrate the phase modulator and to align the phases of the harmonic frequencies. Since the polarization of the summed output is orthogonal to that of the harmonics, a polarizer is used to project the polarization of the two states onto a common axis in order to maximize the heterodyning signal. With five frequency components, four measurements are needed for determining of their relative phases. The setup of the relative phase between each harmonic is shown as Figure 7.

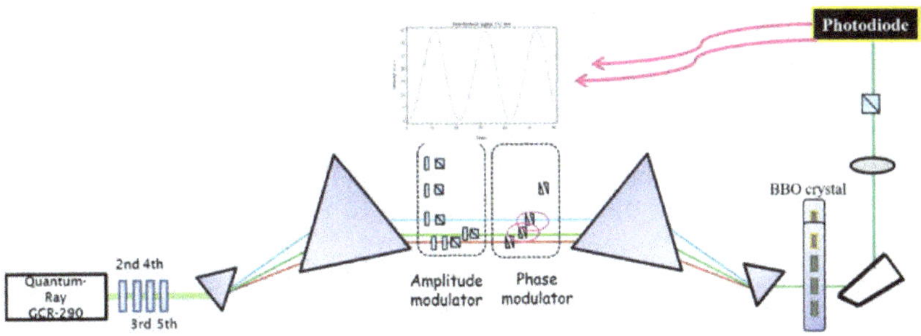

Figure 7. The experimental setup for relative phase measurement. Four BBO crystals were used. Second, third, fourth and fifth indicate the nonlinear crystals in the cascaded generation process.

The flow chart of determining the relative phase of the second harmonic (532 nm) with respect to the fundamental by measuring the interference signal is shown in Figure 8. First, light from the laser system at ω_1 and ω_2 generates the signal at ω_3' through the sum-frequency generation (SFG) in a BBO crystal. By tuning the phase modulator inserted in the beam path of the light at ω_3 from the laser signal, we can introduce a phase difference $\Delta\phi_3$ between the harmonic from the laser system and light of the same frequency from the sum-frequency generation process. For the case of 355 nm light (see Figure 9),

$$\Delta\phi_{355} = \phi'_{355} - \phi_{355} = \phi_{1064} + \phi_{532} + \frac{\pi}{2} - \phi_{355}$$

Or

$$\phi_{355} = 3\phi_{1064} + \pi - \Delta\phi_{532} - \Delta\phi_{355} \tag{3}$$

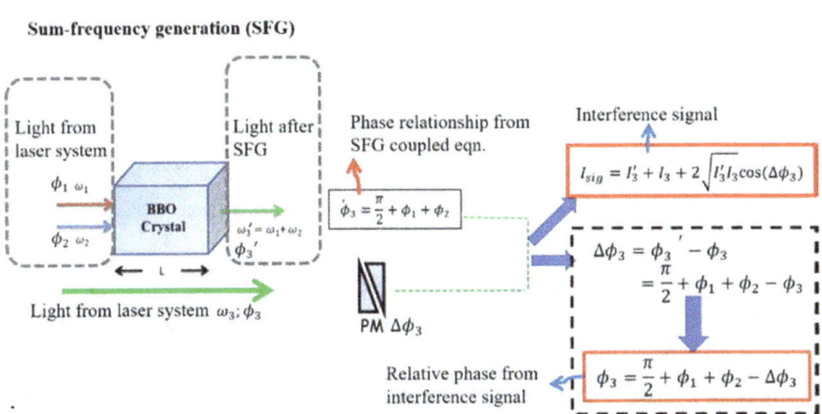

Figure 8. The flow chart for measuring the relative phase through the interference signal.

As the phase modulator is tuned, the interference signal shows the expected sinusoidal behavior (see Figure 9). After the relative phase changes over a few cycles, the scan is stopped (middle of Figure 9). This part of the interference record reflects the phase stability of the system, as the phase and power of the harmonics do vary in practice.

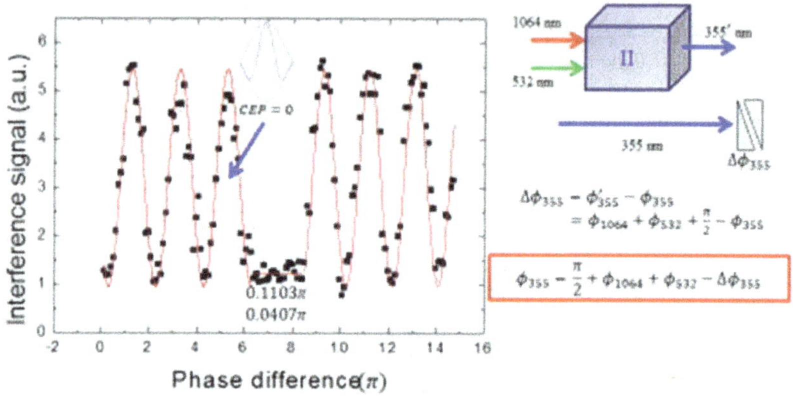

Figure 9. The relative phase between fundamental and the third harmonic is determined as shown on the right of the figure. Left of the signal shows the experimentally measured interference signal. CEP: carrier envelope phase.

The phase stability of the third-harmonic beam at 355 nm is 0.0407 π, while that of the second harmonic is 0.1103 π. It is possible to control the phase modulator such that the interference signal is maintained at certain level, for example, the half, the maximum, and the minimum of the magnitude of the interference signal. For example, we fixed $\Delta\ \phi$ 532 $= \pi / 2$. When the interference signal at 355 nm is at half of the maximum intensity, the phase difference $\Delta\ \phi_{355}$ is 0.5π. According to Eq. (3), the relative phase relationship is $\phi_{355} = 3\ \phi$ 1064 which is the phase-matching condition. The carrier envelope phase of the synthesized wave or CEP is zero. Similarly, if we set the phase difference $\Delta\phi_{355}$ to be 0. $\phi_{355} = 3\ \phi_{1064} + \pi / 2$. Therefore, the CEP of the synthesized waveform is $\pi/2$.

4. WAVEFORM SYNTHESIS AND ITS MEASUREMENT

As the spectral bandwidth of this coherent laser source exceeds two octaves or 32,200 cm^{-1}, conventional methods for ultrafast waveform synthesis is not adequate. We used the shaper-assisted linear correlation method [19] for such a task. This method is particularly suited for diagnostics of multiwave synthesized waveforms.

The basic concept is the use of an effective delta function waveform to retrieve the waveform. To begin with, the output electric field of a coherent multiwave synthesized optical waveform, for example, a mode-locked laser can be expressed as:

$$E_a(t) = \sum_{n=1}^{N} a_n \cos\left(n\omega t + \phi_{an\omega} + \phi_{aCEP}\right) \tag{4}$$

where a_n and $\phi_{an\omega}$ are the amplitude and phase of each component at the frequency $n\omega$, n is a positive integer. ϕ aCEP is the carrier envelope phase. Considering two such waveforms, one is the reference with field E a t above and the target waveform with field $E_b(t)$, given by

$$E_b(t) = \sum_{n=1}^{N} b_n \cos\left(n\omega t + \phi_{bn\omega} + \phi_{bCEP}\right). \tag{5}$$

The interference of the two with a relative temporal delay τ can be described as follows:

$$E_T(t,\tau) = \frac{1}{2}\sum A_n e^{i\left(n\omega t + \phi_{n\omega}\right)} + c.c. \tag{6}$$

$$\text{(where) } A_n = \sqrt{a_{n+}^2 b_n^2 + 2a_n b_n \cos\left(n\omega\tau + \left(\phi_{bn\omega} - \phi_{an\omega}\right) + \phi_{bCEP} - \phi_{aCEP}\right)}$$

$$\phi_n = \cos^{-1}\left[\left(\left(a_n \cos\left(\phi_{an\omega} + \phi_{aCEP}\right) + b_n \cos\left(n\omega\tau + \phi_{bn\omega} + \phi_{bCEP}\right)\right)/A_n\right]$$

A_n and $\phi_{n\omega}$ are the amplitude and phase of the n_{th} Fourier component of the interference signal. The linear cross-correlation function of the reference and target signals with a relative time delay of τ . The time-averaged intensity of E_T is then given by

$$\begin{aligned} I(\tau) &= \frac{1}{T}\int E_T(t,\tau)E_T^*(t,\tau)dt = \frac{1}{4}\sum_n A_n^2 \\ &= \sum \left(a_{n+}^2 b_n^2 + 2a_n b_n \cos\left(n\omega\tau + n\left(\phi_{bn\omega} - \phi_{an\omega}\right) + \phi_{bCEP} - \phi_{aCEP}\right)\right) \end{aligned} \tag{7}$$

If the reference waveform is a transform-limited cosine pulse function of finite duration or a delta function of unity amplitude, that is, $a_n = a_0$, phase $\phi_{an\omega} = 0$, $\phi_{aCEP} = 0$

$$I(\tau) = \sum \left(a_{0+}^2 b_n^2\right) + 2a_0 b_n \cos\left(n\omega\tau + \phi_{bn\omega} + \phi_{bCEP}\right) \tag{8}$$

That is, the time-varying part of $I(\tau)$ is directly proportional to the target field, Eb(t) (see Eq. (6)). If the reference pulse and target one are delta and square pulse, Eq. (5) can be written as

$$
\begin{aligned}
E(t,\tau) &= E_\delta(t) + E_{square}(t,\tau) \\
&= A_\delta \sum_n e^{i(\omega_n t - k_n d)} + B_{squ} \sum_{n=1,3,5,..} \frac{2}{n\pi} e^{i\left(\omega_n(t-\tau) - k_n d - \frac{\pi}{2}\right)} \\
&= \sum_n e^{i\omega_n t} A_n'(\tau) e^{i\varphi_n'(\tau)}
\end{aligned} \tag{9}
$$

where

$$
A_n' = \sqrt{A_\delta^2 + \left(\frac{B_{squ}}{n}\right)^2 + 2A_\delta \frac{B_{squ}}{n}\cos\left(\omega_n\tau - \frac{\pi}{2}\right)},
$$

$$
\varphi_n' = \tan^{-1}\left[\frac{-A_\delta \sin(k_n d) + \frac{B_{squ}}{n}\sin\left(\omega_n\tau - \frac{\pi}{2} - k_n d\right)}{A_\delta \cos(k_n d) + \frac{B_{squ}}{n}\cos\left(\omega_n\tau - \frac{\pi}{2} - k_n d\right)}\right] \quad \text{for n} = 1, 3, 5.. \tag{10}
$$

$$
A_n' = A_\delta, \quad \varphi_n' = -k_n d \qquad \text{for n} = 2, 4, 6.... \tag{11}
$$

The linear cross-correlation measurement can be performed using any interferometric arrangement, for example, a Michelson interferometer. Equivalently, it can be conducted by adjusting the amplitudes and phases of the frequency components of the waveform. The experimental setup is shown in Figure 10. A thermal pile power meter, which can detect light from the fundamental (λ = 1064 nm) to the fifth harmonic (λ = 213 nm) of the laser system.

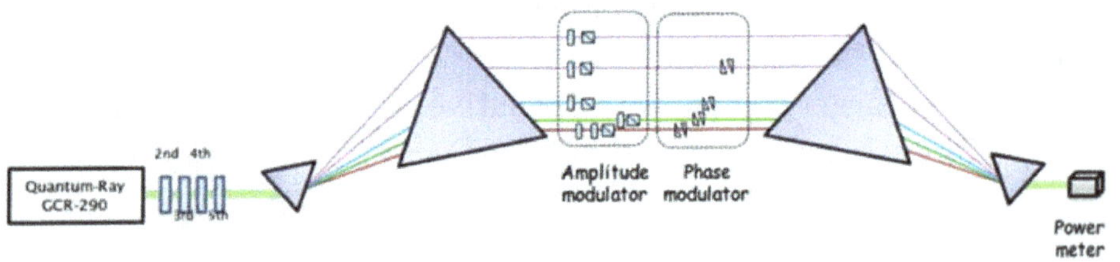

Figure 10. The experimental arrangement for linear cross-correlation measurement of the synthesized waveform. Second, third, fourth and fifth indicate the nonlinear crystals that generated the cascaded harmonics.

We have shown previously that it is possible to synthesize attosecond pulse train and arbitrary waveforms using this approach [11]. For example, Figure 11(a) shows the synthesized square waveform by the fundamental, second and third harmonics of the Nd:YAG laser. The normalized amplitudes of the harmonics are respectively, 1, 0 and 1/3. Figure 11(b) shows the synthesized sawtooth waveform by the fundamental through the fourth harmonics of the Nd: YAG laser.

The normalized amplitudes of the fundamental and harmonics are respectively, 1, 1/2, 1/3 and 1/4. The measured waveforms are in good agreement with the theoretical estimates (solid curves in Figure 11). Although we just used three of four waves in this experiment, the synthesized waveforms already reproduce these familiar mathematical functions.

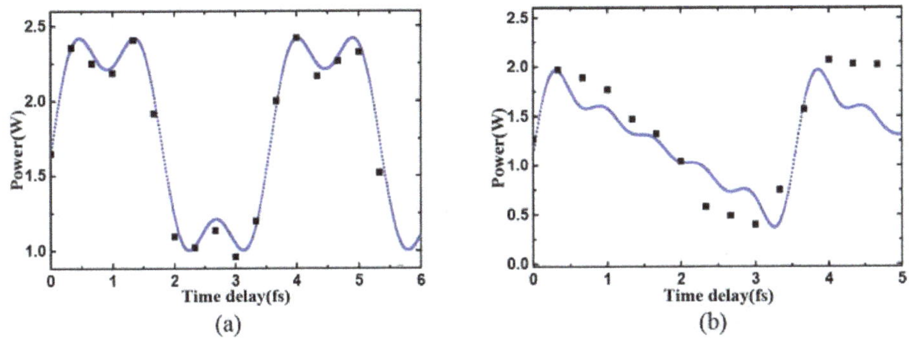

Figure 11. (a) Synthesis of a square waveform with the fundamental, second and third harmonics of the Nd:YAG laser. (b) Synthesis of a sawtooth waveform with the fundamental, second, third and fourth harmonics of the Nd:YAG laser. The solid squares are experimental data. The blue curves are theoretical curves. (reproduced with permission from [11]).

Our laser system can generate fundamental through the fifth harmonics with pulse energies of 380, 178, 70, 41, and 22 mJ. If these can be fully utilized, the synthesized transform-limited pulse will exhibit a temporal FWHM of 480 attoseconds. The intensity envelope will be just 700 attoseconds. The intensity of each attosecond pulse will exceed 10^{14} W/cm^2 when it is focused to a spot size of 20 μm. Interesting nonlinear effects in materials would be induced by such high-power vibrations. Opportunities for innovative laser processing should emerge. These will be addressed in the second section of this work.

5. CONCISE SUMMARY

A novel high-power attosecond light source was proposed and subsequently demonstrated through frequency synthesis. The Q-switched Nd:YAG laser system is comprised of a narrow-band transform-limited high-power laser and its second (λ = 532 nm) through fifth harmonics (λ = 213 nm). The laser system was engineered to ensure that the cascaded harmonics spatially overlap and co-propagate to the far fields. Thus, the spectral bandwidth of this coherent laser source surpasses two octaves or 32,200 cm^{-1}.The amplitude and phase of the comb, which is composed of the five frequency components, can be independently regulated. This method can generate sub-femtosecond (360 attosecond) pulses with carrier-envelope phase (CEP) control that are sub-single-cycle (~ 0.37 cycle). With a focused spot size of 20 μm, the maximal intensity of each pulse exceeds 1014 W/cm^2. Additionally, it is feasible to generate optical waveforms of any shape, such as a square wave. The waveform that has been synthesized is stable for a minimum of thousands of nanoseconds.

AT A GLANCE

We introduce a novel approach to the synthesis of arbitrary waveforms and high-power attosecond pulses through multicolor synthesis. The multicolor laser system's full bandwidth, which extends over two octaves and reaches 37,600 cm^{-1}, can be employed to produce sub-femtosecond (360 attosecond) pulses with carrier-envelope phase (CEP) control that are sub-single-cycle (approximately 0.37 cycle). The results indicate that the approach is promising for the generation of relatively high-power attosecond pulses in the optical region. The laser system's design and diagnostics are illustrated in this chapter. In the subsequent chapter, we present a selection of the novel source's applications, including coherently controlled harmonic generation and phase-sensitive 2-color ablation of copper and stainless steel using this multi-color laser system.

REFERENCES

1. Helmer M. Nature Milestones: Photons, Milestone. Attosecond science 1 May, 2010;Vol. 22; 2001. doi:10.1038/nmat2659
2. Krausz F, Ivanov M. Reviews of Modern Physics. 2009;81:163
3. Goulielmakis E, Yakovlev VS, Cavalieri AL, Uiberacker M, Pervak V, Apolonski A, Kienberger R, Kleineberg U, Krausz F. Science. 2007;317:769-775
4. Cavalieri AL, Müller N, Uphues T, Yakovlev VS, Baltus caronka1 A, Horvath B, Schmidt B, Blümel L, Holzwarth R, Hendel S, Drescher M, Kleineberg U, Echenique PM, Kienberger R, Krausz F, Heinzmann U. Nature. 2007;449:1029-1032

5. Liu X, Du D, Mourou G. IEEE Journal of Quantum Electronics. 1997;33:1706-1716

6. Lewenstein M, Balcou P, Ivanov MY, L'Huillier A, Corkum PB. Physical Review A. 1994;49(3):2117-2131

7. Midorikawa K, Nabekawa Y, Suda A. XUV multiphoton processes with intense high-order harmonics. Progress in Quantum Electronics. 2008;32(2):43-88

8. Chen W-J, Hsieh Z-M, Huang SW, Su H-Y, Lai C-J, Tang T-T, Lin C-H, Lee C-K, Pan R-P, Pan C-L, Kung AH. Physical Review Letters. 2008;100:163906

9. Hsieh Z-M, Lai C-J, Chan H-S, Sih-Ying W, Lee C-K, Chen W-J, Pan C-L, Yee F-G, Kung AH. Physical Review Letters. 2009;102:213902

10. Chen W-J, Lee C-K, Pan C-L. Paper FWE6, Presented at the Frontiers in Optics 2010/Laser Science XXVI, OSA Technical Digest (CD); Rochester, New York, USA; Oct. 2010. pp. 24-28

11. Chen W-J, Wang H-Z, Lin R-Y, Lee C-K, Pan C-L. Laser Physics Letters. 2012;9(3):212

12. Chen W-J, Lin R-Y, Chen W-F, Lee C-K, Pan C-L. Laser Physics Letters. 2013;10:065401

13. Goodman JW. Introduction to Fourier Optics. New York: McGraw-Hill; 2002

14. Chen W-J, Kung AH. Optics Letters. 2005;30:2608

15. Huang SW, Chen WJ, Kung AH. Physical Review A. 2006;74:063825

16. Hansch TW. Optics Communications. 1990;80:71-75

17. Mukai T, Wynands R, Hansch TW. Optics Communications. 1993;95:71-76

18. Wang Z, Bovik AC. IEEE Signal Processing Letters. 2002;9:81-84

19. Chan H-S, Hsieh Z-M, Liang W-H, Kung AH, Lee C-K, Lai C-J, Pan R-P, Peng L-H. Science. 2011;331:1165-1168

CHAPTER-8

HIGH-POWER ATTOSECOND PULSE GENERATION AND ITS APPLICATIONS USING A FREQUENCY-SYNTHESIZED APPROACH

1. INTRODUCTION: AN OVERVIEW

In the first section of this study [1], we presented a novel high-power laser system capable of producing attosecond light pulses and arbitrary waveforms through frequency synthesis. The laser system has the capability to produce up to five collinear beams with regulated amplitude and phase. These beams have different wavelengths, ranging from the fundamental output of the Nd:YAG laser at 1064 nm, to its second harmonic at 532 nm, and up to the fifth harmonic at 213 nm. Using this method, it is possible to generate sub-single-cycle pulses with a duration of approximately 0.37 cycles and a duration of 360 attoseconds. These pulses may also be controlled in terms of their carrier-envelope phase (CEP). Each pulse has a peak intensity that surpasses 10^{14} W/cm^2, and the concentrated spot size is 20 μm. The study also showcased stable square and saw-tooth waveforms [2].

The precise manipulation of nonlinear optical processes, such as harmonic production, through the control of laser field waveforms, is crucial for advancing both fundamental scientific understanding and practical technology applications. In our previous study, we examined how the respective phases and intensities of the two-color pump electric fields (1064 and 532 nm) affect the third-order nonlinear frequency conversion process in argon [2]. Experimental evidence demonstrated that the third-harmonic (TH) signal exhibits regular oscillations in accordance with the relative phases of the two-color driving laser fields. The data align well with a perturbative nonlinear optical study of the third-harmonic (TH) signal. This analysis includes the direct third-harmonic-generation (THG), four-wave mixing (FWM), and the interference of these two processes.

As a continuation of our research, we have investigated the production of harmonics using a three-color synthetic waveform in inert gas systems. In Section 2, we shall demonstrate the physics involved by analyzing the instance of fourth-harmonic generation (FHG). There was a noticeable increase in the THz signal when the 532 nm beam was present. This study demonstrates that the plasma produced by the ionization process during the contact between laser and materials has a substantial impact on the amplification of the TH signal. We additionally showcased the phase-sensitive two-color ablation of copper and stainless steel. Our findings indicate that hole drilling is a more effective method for optimal waveforms.

2. NONLINEAR FREQUENCY CONVERSION BY COHERENTLY CONTROLLED THREE-COLOR EXCITATION OF INERT GASES

This section explores the utilization of three-color laser fields as a means to produce harmonic signals in an isotropic medium, such as inert gasses. By utilizing a three-color pump and focusing just on the lowest order nonlinear processes in isotropic systems, specifically the third-order nonlinear process, it is possible to generate harmonics ranging from the 4th to the 9th of the laser fundamental output. It is anticipated that there would be a multitude of nonlinear effects and intricate quantum interference events. This information is presented concisely in Table 1.

By employing perturbative nonlinear formulism, we initially developed the comprehensive formula for the harmonic electric field, together with its corresponding intensity. The coherent effect is observed when two frequency conversion circuits interfere with each other. In the following, we will utilize the situation of FHG as an exemplification of the anticipated physical events.

Table 1. Third-order nonlinear process ($\omega_n = \omega_i + \omega_j + \omega_k$, $\omega_n = \omega_i + \omega_j - \omega_k$, n = 4–9, i, j, k = 1–3) that can contribute to the generation of 4th to 9th harmonics of the laser fundamental output of a three-color field (the fundamental ω_1, second harmonic $\omega_2 = 2\omega_1$ and third harmonic $\omega_3 = 3\omega_1$).

Harmonic generation	Processes
Fourth	$\omega_1+\omega_1+\omega_2\rightarrow\omega_4$
	$-\omega_1+\omega_2+\omega_3\rightarrow\omega_4$
	$-\omega_2+\omega_3+\omega_3\rightarrow\omega_4$
Fifth	$\omega_1+\omega_2+\omega_2\rightarrow\omega_5$
	$\omega_1+\omega_1+\omega_3\rightarrow\omega_5$
	$-\omega_1+\omega_3+\omega_3\rightarrow\omega_5$
Sixth	$\omega_2+\omega_2+\omega_2\rightarrow\omega_6$
	$\omega_1+\omega_2+\omega_3\rightarrow\omega_6$
Seventh	$\omega_2+\omega_2+\omega_3\rightarrow\omega_7$
	$\omega_1+\omega_3+\omega_3\rightarrow\omega_7$
Eighth	$\omega_2+\omega_3+\omega_3\rightarrow\omega_8$
Ninth	$\omega_3+\omega_3+\omega_3\rightarrow\omega_9$

By using a three-color field consisting of the fundamental frequency ω1, the second harmonic ω2 (which is twice the frequency of ω1), and the third harmonic ω3 (which is three times the frequency of ω1), it is possible to generate a fourth-harmonic signal. This is achieved by three nonlinear optical processes: the sum of ω1, ω1, and ω2 resulting in ω4, the difference between -ω1, ω2, and ω3 resulting in ω4, and the difference between -ω2, ω3, and ω3 resulting in ω4. The modulation of the conversion efficiency for the fourth-harmonic signal can be achieved through the interference between any two of the three Four-Wave Mixing (FWM) processes. As the relative phase between ω1, ω2, and ω3 changes, different combinations of three sinusoidal modulations cause interference in the output intensity of the fourth harmonic at frequency ω4. In addition, we will demonstrate that the relative magnitude of the primary, secondary, and tertiary driving laser fields affects the fourth-harmonic signal.

We assume plane-waves propagating in the +z direction. The three-color field can be represented as:

$$\tilde{E}(z,t) = \frac{1}{2}\left[E_1 e^{i(k_1 z-\omega_1 t+\varphi_1)} + E_2 e^{i(k_2 z-\omega_2 t+\varphi_2)} + E_1 e^{i(k_3 z-\omega_3 t+\varphi_1)} + c.c.\right] \quad (1)$$

where φ_1, φ_2 and φ_3 are the modulated phases of the three colors, respectively. As a source, the nonlinear polarization term in the medium induced by the three-color fields will generate several new frequency components. If we only consider third-order nonlinear optical processes only, assuming no pump depletion, the electric field of the fourth harmonic can then be rewritten as

$$E_4(z) = E_{4I}(z) + E_{4II}(z) + E_{4III}(z), \quad (2)$$

where

$$E_{4I}(z) = i\frac{12\pi^2}{n_4\lambda_1} N\chi^{(3)}_{4I} E_1^2 E_2 L \sin c\left(\frac{\Delta k_{4I}L}{2}\right) e^{i\Delta k_{4I}L} e^{i(2\varphi_1+\varphi_2)}. \quad (3)$$

$$E_{4II}(z) = i\frac{24\pi^2}{n_4\lambda_1} N\chi^{(3)}_{4II} E_1^* E_2 E_3 L \sin c\left(\frac{\Delta k_{4II}L}{2}\right) e^{i\Delta k_{4II}L} e^{i(-\varphi_1+\varphi_2+\varphi_3)}, \quad (4)$$

$$E_{4III}(z) = i\frac{12\pi^2}{n_4\lambda_1} N\chi^{(3)}_{4III} E_2^* E_3^2 L \sin c\left(\frac{\Delta k_{4III}L}{2}\right) e^{i\Delta k_{4III}L} e^{i(-\varphi_2+2\varphi_3)}, \quad (5)$$

With the phase or wave-vector mismatch given by $\Delta k_{4I} = k_4 - 2k_1 - k_2$, $\Delta k4II = k_4 + k_1 - k_2 - k_3$ and $\Delta k_{4III} = k_4 + k_2 - 2k_3$. In this section, the symbol "I","II", "III" represent the three possible four-wave mixing (FWM) processes with corresponding nonlinear susceptibilities: $\chi^{(3)}{}_{4I} = \chi^{(3)}{}_{(}\omega_4; \omega_1, \omega_1, \omega_{2)}$, $\chi^{(3)}{}_{4II} = \chi^{(3)}{}_{(}\omega_4; -\omega_1, \omega_2, \omega_{3)}$, $\chi^{(3)}{}_{4III} = \chi^{(3)}{}_{(}\omega_4; -\omega_2, \omega_3, \omega_3)$, respectively. L stands for the nonlinear medium length. The intensity of the fourth-harmonic signal can then be written as

$$I_4(z) = \frac{cn_4}{8\pi}\left\langle \left|\tilde{E}_{4I}(z,t) + \tilde{E}_{4II}(z,t) + \tilde{E}_{4III}(z,t)\right|^2 \right\rangle \quad (6)$$

or

$$I_4(z) = \frac{9216\pi^6 N^2 L^2}{c^2 \lambda_1^2 n_4} \left\{ \begin{array}{l} \frac{1}{n_1^2 n_2}\left(\chi_{4I}^{(3)}\right)^2 I_1^2 I_2 sinc^2\left(\frac{\Delta k_{4I} L}{2}\right) \\ +\frac{4}{n_1 n_2 n_3}\left(\chi_{4II}^{(3)}\right)^2 I_1 I_2 I_3 sinc^2\left(\frac{\Delta k_{4II} L}{2}\right) \\ +\frac{1}{n_2 n_3^2}\left(\chi_{4III}^{(3)}\right)^2 I_2 I_3^2 sinc^2\left(\frac{\Delta k_{4III} L}{2}\right) \\ +\frac{2}{\sqrt{n_1 n_3 n_1 n_3}} \chi_{4I}^{(3)} \chi_{4II}^{(3)} \sqrt{I_1 I_3} I_1 I_2 sinc\left(\frac{\Delta k_{4I} L}{2}\right) \\ \times sinc\left(\frac{\Delta k_{4II} L}{2}\right) \cos\left(\frac{\Delta k_{4I} L}{2} - \frac{\Delta k_{4II} L}{2} + 3\varphi_1 - \varphi_3\right) \\ +\frac{2}{\sqrt{n_1 n_3 n_2 n_3}} \chi_{4II}^{(3)} \chi_{4III}^{(3)} \sqrt{I_1 I_3} I_2 I_3 sinc\left(\frac{\Delta k_{4II} L}{2}\right) \\ \times sinc\left(\frac{\Delta k_{4III} L}{2}\right) \times \cos\left(\frac{\Delta k_{4II} L}{2} - \frac{\Delta k_{4III} L}{2} - \varphi_1 + 2\varphi_2 - \varphi_3\right) \\ +\frac{1}{n_1 n_2 n_3} \chi_{4III}^{(3)} \chi_{4I}^{(3)} I_1 I_2 I_3 sinc\left(\frac{\Delta k_{4III} L}{2}\right) \\ \times sinc\left(\frac{\Delta k_{4I} L}{2}\right) \cos\left(\frac{\Delta k_{4III} L}{2} - \frac{\Delta k_{4I} L}{2} - 2\varphi_1 - 2\varphi_2 + 2\varphi_3\right) \end{array} \right\} \quad (7$$

In Eq. (7), the first, second and third terms are the three FWM processes, I, II and III, respectively. The last three terms are cross-terms due the interference of the optical fields generated by FWM processes I and II, II and III and III and I, in that order.

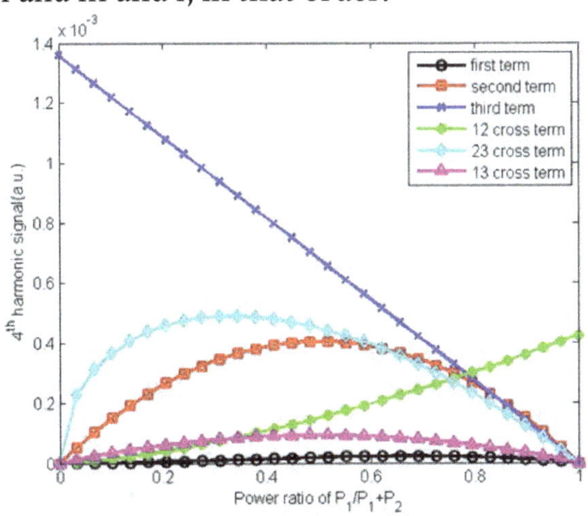

Figure 1. The fourth-harmonic signal versus the power ratio $P_1/P_1 + P_2$. Contributions by the three FWM processes and the cross-terms are shown as different colors.

In the simulation, we used three-color laser fields (the fundamental, second harmonic, and third harmonic of the Nd:YAG laser) to generate fourth-harmonic signal in gaseous argon. For simplicity, we further assumed that the phase mismatch for all of FWM processes is equal and negligible. Further, the fundamental and second-harmonic power is the same and their sum is normalized.

In Figure 1 we show the fourth-harmonic signal as function of the power ratio of the fundamental beam and that of the fundamental and second harmonic combined. The third-harmonic beam is held constant. Examining Figure 1, one can see clearly that much higher conversion efficiency of the fourth-harmonic signal would be generated if the normalized power ratio is around 0.8.

The dependence of the fourth-harmonic signal on the phase of the fundamental beam is shown in Figure 2. Clearly, the modulation is more complex than the two-color case.

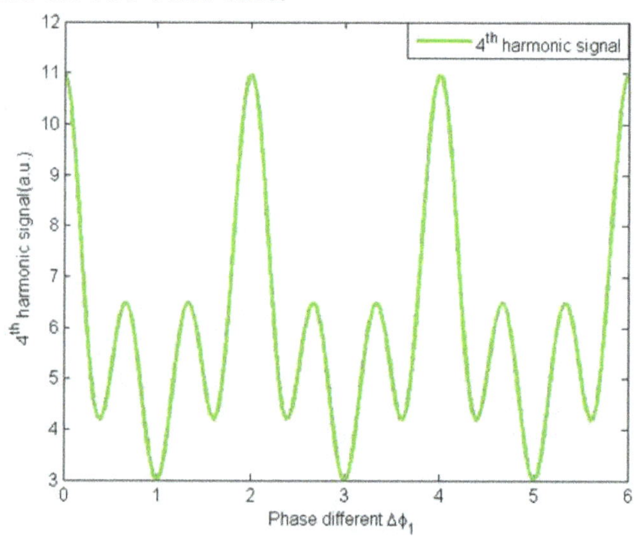

Figure 2. The dependence of the fourth-harmonic signal on the phase of the fundamental beam.

3. THIRD-HARMONIC GENERATION BY COHERENTLY CONTROLLED TWO-COLOR EXCITATION OF INERT GASES: PLASMA EFFECT

With two-color excitation, the third-harmonic signal is contributed by the direct THG ($\omega_3 = \omega_1 + \omega_1 + \omega_1$) and four-wave mixing (FWM, $\omega_3 = \omega_2 + \omega_2 - \omega_1$) processes and a cross term of the two. As the relative phase between ω_1 and ω_2 varies, a sinusoidal modulation in output intensity at frequency ω_3 is expected and was demonstrated in our previous work [2]. In intense laser field, plasma can be generated through the ionization of gases. Optical harmonic generation in plasmas has been studied for a long time. Recently, significant enhancement of the third-harmonic emission in plasma has been reported by Suntsov et al. [3]. More than two-order-of-magnitude increase of the efficiency of third-harmonic generation occurs due to the plasma-enhanced third-order susceptibility [5]. More specifically, the presence of charged species (free electrons and ions) can effectively increase the third-order nonlinear optical susceptibility [4, 5]. This indicates that the susceptibility can be expressed as a function of the plasma density *Ne* induced by laser field. Additionally, the refractive index of the target, for example, gases or solids, is also changed in the presence of the plasma. The wave-vector mismatch Δk, in plasma, between the fundamental and the third-harmonic signal can be derived by using the Drude model. Enhanced third-harmonic signal that eventually saturates at higher plasma density was predicted [6, 7]. In this chapter, we observed more than ten orders of magnitude enhancement of third-harmonic generation in argon plasma by employing the fundamental (1064 nm) and second-harmonic (532 nm) fields of an injection-seeded Q-switched Nd:YAG laser. Under the assumption that susceptibility and wave-vector mismatch depend on the plasma density, we show that plasma plays a significant role in the third-harmonic signal by an analysis based on the formulism of perturbative nonlinear optics. Significant enhancement of the TH signal is caused by the plasma-enhanced susceptibility of the dominant four-wave mixing process. When the plasma density is high enough, the TH signal becomes saturated and drops primarily due to the detrimental effect of the wave-vector mismatch.

The experimental setup for studying the effect of plasma formation on generation of third-harmonic signal by phase-controlled two-color excitation is shown in Figure 3. It is a simplified version of the multicolor laser system described in part I of this work and our previous papers [2, 3]. To reiterate, we employed a Q-switched Nd:YAG laser system (Spectra Physics GCR Pro-290) that generates intense 1064 nm pulses with a pulse duration of 10 ns (FWHM) and a line width of $<0.003\ cm^{-1}$.

The laser pulse repetition rate is 10 Hz, and the maximum pulse energy is 1.9 J/pulse. The second-harmonic (532 nm) beam was generated by using the nonlinear optical crystal KD*P (type I phase matching). The maximum pulse energy of the second-harmonic signal is around 1 J/pulse. The fundamental and second-harmonic pulses propagate collinearly with a fixed relative phase. This two-color laser beams are separated by a prism pair into two arms. A power tunable two-color system can be generated with two amplitude modulators for each arm. The relative phase and amplitudes of these two-color laser fields can be timed independently by amplitude and phase modulators. The fundamental and second-harmonic beams are first angularly separated and then made parallel by a pair of prisms. With the desired amplitude ratio and relative phase, the two-color laser fields are recombined with an identical pair of prisms and then focused into a vacuum chamber filled with argon (10 Torr) by a 10-cm-focal lens to generate the third-harmonic (355 nm) signal. To overlap two foci of the fundamental and second-harmonic beam, the dispersion of the lens is compensated by a telescope in the fundamental arm. The third-harmonic generation is filtered by a monochromator (VM-502, Acton Research) and detected by a photomultiplier tubes (R11568, Hamamatsu).

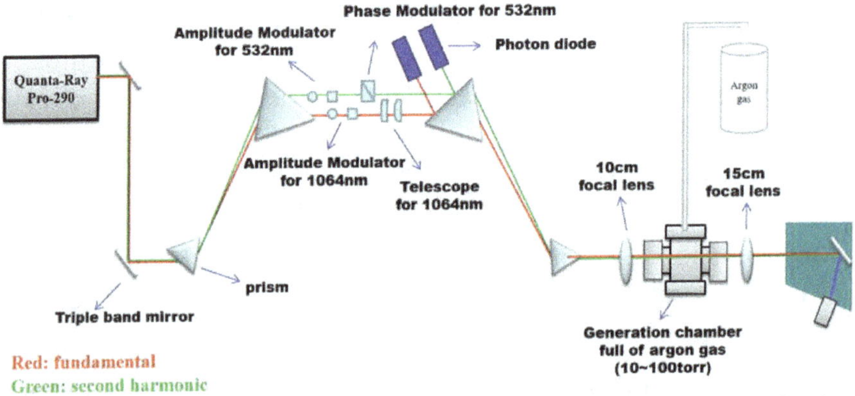

Figure 3. The experimental setup for studying the effect of plasma formation on generation of third-harmonic signal by phase-controlled two-color excitation.

With excitation by the two-color field (the fundamental ω_1 and second harmonic ω_2) of the Nd:YAG laser, the third-harmonic signal can be generated by two optical processes, i.e., $\omega_1 + \omega_1 + \omega_1 = \omega_3$ and $-\omega_1 + \omega_2 + \omega_2 = \omega_3$. We assume plane-waves propagating in the +z direction.

The theoretical formulism is similar to the three-color case in Section 1. In the slow-varying envelope approximation and assume no pump depletion, the TH field can be written as

$$\widehat{E}_3(z) = \widehat{E}_{3I}(z) + \widehat{E}_{3II}(z) = \left\{ \begin{array}{c} i\dfrac{3\pi^2}{n_3\lambda_1}\chi_I^{(3)}E_1^3L\mathrm{sinc}\left(\dfrac{\Delta k_{13}L}{2}\right)e^{i\frac{\Delta k_{13}L}{2}}e^{i3\varphi_1} \\ +i\dfrac{9\pi^2}{n_3\lambda_1}\chi_{II}^{(3)}E_1^*E_2^2L\mathrm{sinc}\left(\dfrac{\Delta k_{213}L}{2}\right)e^{i\frac{\Delta k_{213}L}{2}}e^{i(2\varphi_2-\varphi_1)} \end{array} \right\} \tag{8}$$

where the subscripts "I" and "II" denote the two nonlinear processes, namely the direct THG and FWM; $\chi^{(3)}$ is the third-order nonlinear susceptibility, L is the length of the nonlinear material; φ 1 and φ 2 are the phases of the fundamental and second-harmonic beams, respectively. $\Delta k_{13} = 3k_1 - k_3$ and $\Delta k_{213} = 2k_2 - k_1 - k_3$ are wave-vector mismatch due to dispersion in the gaseous media. The refractive index of the gas can be calculated by using Sellmeier equation,

$$\begin{aligned} n(\lambda) - 1 &= (n-1)_{lines} + (n-1)_{cont} \\ &= \frac{N_g r_e}{2\pi}\sum_i \frac{f_i}{\lambda_i^{-2} - \lambda^{-2}} + \frac{N_g}{2\pi^2}\int \frac{\sigma d\overline{v}_i}{\overline{v}_i^2 - \overline{v}^2}. \end{aligned} \tag{9}$$

In Eq. (9), $N_g = P/k_BT$ is the gas density related to the pressure of the gas by ideal gas law N_g in which k_B is Planck constant; r_e is classical electron radius. The first term or $(n - 1)_{lines}$ refers to the contribution by discrete energy levels of the atom while the second term or $(n - 1)_{cont}$ is that by the continuum states. For the oscillator strengths f of argon, we used those listed in Ref. [8]. In addition, we take the photoionization cross-section from Ref. [9]. The intensity of TH signal, therefore, can be written as

$$\begin{aligned} I_3(\varphi) &= \frac{cn_3}{8\pi}\left\langle \left|\widehat{E}_{3I}(z,t) + \widehat{E}_{3II}(z,t)\right|^2 \right\rangle = I_{3I} + I_{3II} + I_{3III} \\ &= \frac{576\pi^6L^2I_1}{c^2n_3n_1\lambda_1^2}\left\{ \begin{array}{c} \dfrac{1}{n_1^2}\left(\chi_I^{(3)}\right)^2 I_1^2\mathrm{sinc}^2\left(\dfrac{\Delta k_{13}L}{2}\right) \\ +\dfrac{9}{n_2^2}\left(\chi_{II}^{(3)}\right)^2 I_2^2\mathrm{sinc}^2\left(\dfrac{\Delta k_{213}L}{2}\right) \\ +\dfrac{6}{n_1n_2}\chi_I^{(3)}\chi_{II}^{(3)}I_1I_2\mathrm{sinc}\left(\dfrac{\Delta k_{13}L}{2}\right)\mathrm{sinc}\left(\dfrac{\Delta k_{213}L}{2}\right) \\ \times\cos\left(\dfrac{\Delta k_{13}L}{2} - \dfrac{\Delta k_{213}L}{2} + 4\varphi_1 - 2\varphi_2\right) \end{array} \right\} \end{aligned} \tag{10}$$

In Eq. (10), the first, second and third term corresponds to THG. FWM and a cross-term due to the interference of the former two processes. For the sake of simplicity, we can set $\phi_1 = 0$. Therefore, $\Delta\phi = \phi_2 - 2\phi_1 = \phi_2$ is the relative phase between ω_1 and ω_2. In media with normal dispersion, for example, the non-resonant excitation of room-temperature argon gas, the relative magnitude of the wave vectors is $k_1 < k_2 < k_3$.

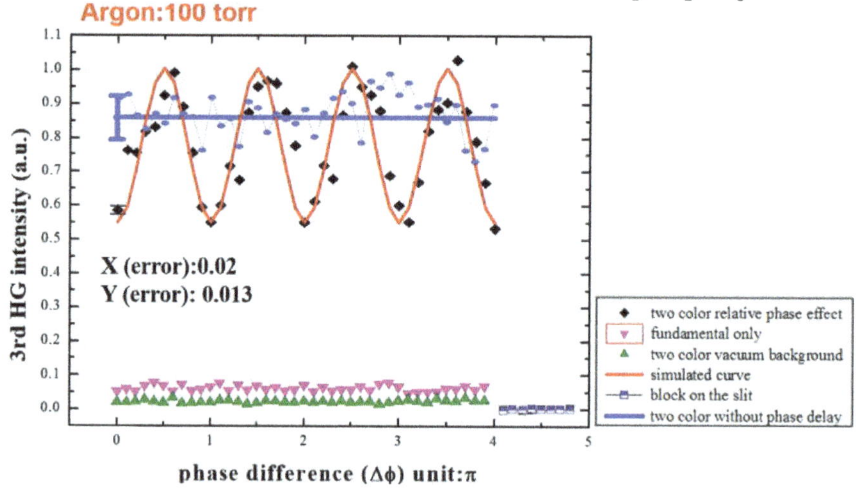

Figure 4. Typical trace of the TH signal plotted as a function of the relative phase between the fundamental and second-harmonic beams. The system noise level corresponding to situation in which the slit of the monochrometer was closed is also shown.

Accordingly, the phase mismatch, Δ k 13 > Δ k 123 ≈ 0 , is negligible. A sinusoidal dependence of the TH signal on the relative phase is thus expected. An example is shown in Figure 4. The pulse energy of the 1064 nm and 532 nm beams were 70 and 1 mJ, respectively. The pressure of the argon gas was 100 Torr. As the beams are slightly elliptical, we measured the TH signal in two transverse directions. The percentile errors in the X- and Y-directions are shown. Note that the TH signal is very weak if only the fundamental beam is used for excitation.

It was found that the TH signal can be enhanced by more than one order of magnitude with two-color excitation. In Table 2, we summarize the phase modulation and enhancement of the TH signal with two-color excitation for several ratios of fundamental and second-harmonic pulse energies. The fluctuations of the TH signal when the relative phase of the fundamental and second-harmonic beams is a constant is also shown.

Table 2. The phase modulation and enhancement of the TH signal with two-color excitation for several ratios of fundamental and second-harmonic pulse energies.

Excitation Source	The modulation of phase or contrast (Normalized)	The fluctuation in TH power without phase delay (normalized)	Enhancement ratio (two-colour/one color)
(1) 1064 (**70mJ/pulse**) +532 (**1mJ/pulse**)	0.45	0.065	9.02~16.38
(2) 1064 (**110mJ/pulse**) +532 (**1mJ/pulse**)	0.6258	0.166	2.28~5.52
(3) 1064 (**70mJ/pulse**) +532 (**20mJ/pulse**)	0.3446	0.1457	13.41~20.58

We observed the enhancement of the TH signal is substantial for two-color excitation. Plasma emission was found to be visible to the naked eye in such cases. It is reasonable to assume that laser-induced ionization in the inert gases, for example, argon. With our experimental conditions, the ionization process is in the multiphoton ionization regime, which occurs when an atoms simultaneously absorbing several photons. The multiphoton ionization rate $w(\omega,F)$ can be calculated by the Perelomov-Popov-Terent'ev (PPT) model, where F is the laser fluence. The rate is a function of the laser oscillation frequency and laser field strength. For the two-color case, we assume an effective frequency which is calculated from the power distribution of laser frequency to describe the influence of the two-color electric field on the ionization rate.

$$\omega_{eff} = \frac{\int_0^\infty \omega |E(\omega)|^2 d\omega}{\int_0^\infty |E(\omega)|^2 d\omega} \tag{11}$$

Besides, for our nanosecond pulse, there are several million cycles inside the pulse envelope for fundamental beam in the near infrared. The cycle-averaged ionization rate is thus used in this work [10]. That is,

$$\overline{w}_{PPT}(F_a) = \frac{1}{T_0}\int_0^{T_0} w_{PPT}(t)dt \tag{12}$$

or

$$\overline{w}_{PPT}(F_a) = \sqrt{\frac{2}{\pi}}\sqrt{\frac{3F_a}{2F_0}}w_{PPT}(F_a) \tag{13}$$

The ionization probability of the atoms by the laser pulse can be calculated by solving the rate equation

$$p = 1 - e^{-\int_{-\infty}^{\infty} w_{PPT}(t)dt}. \tag{14}$$

This allows us to calculate the plasma density in terms of the density of the neutral gas.

$$N_e = p \times N_g \tag{15}$$

The step-like pattern shown in Figure 5 for the ionization probability is a result of the rise in effective frequency with an increase in the number of second-harmonic photons. Specifically, additional absorption processes take place when the pulse's effective photon energy approaches the ionization process threshold.

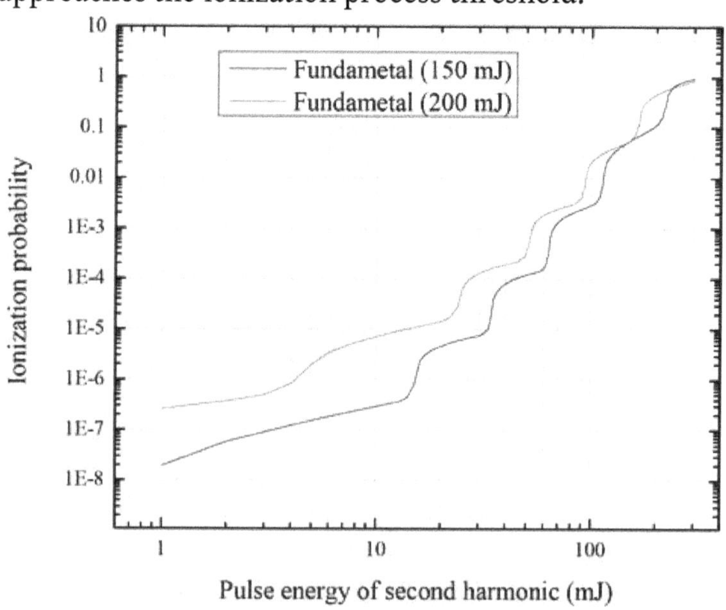

Figure 5. Ionization probability calculated by PPT model for argon at 100 Torr, excited by two-color field with fundamental pulse energies at 150 mJ and 200 mJ.

Now, we consider influence of the plasma on the third-harmonic signal. We assume that the third-order optical susceptibility is a sum of the susceptibilities for the neutral and ionized gas atoms.

$$\chi^{(3)}_{I,p} = \chi^{(3)}_{I,g} + \gamma_{I,p} N_e \tag{16}$$

$$\chi^{(3)}_{II,p} = \chi^{(3)}_{II,g} + \gamma_{II,p} N_e \tag{17}$$

In the above two equations, the ratios $\gamma_{I,p}$ and $\gamma_{II,p}$ are values determined by the experiment. Here, we assume $\gamma_{I,p} = \gamma_{II,p} = 4 \times 10^{-49}$ and $\chi^{(3)}_{I,g} = \chi^{(3)}_{II,g} = \chi^{(3)}_{II,Ar} = 3.8 \times 10^{-26}$ m^2/V^2 [11]. Additionally, we note that the refractive index of the media would also be changed when the plasma is generated. This can be calculated, in the first approximation, by using the Drude model.

$$n_{\omega,p} = n_{\omega,g} - \frac{N_e}{2N_c} \tag{18}$$

where $N_c = \varepsilon_0 m_e \omega^2/e^2$ is the critical plasma density when the laser and plasma frequencies are equal. The subscripts p and g represent plasma and neutral gas, respectively. For argon, $n_{\omega,g} = n_{\omega,Ar}$. The wave-vector mismatch becomes $\Delta k_{13,p} = 3k_{1,p} - k_{3,p}$ and $\Delta k_{213,p} = 2k_{2,p} - k_{1,p} - k_{3,p}$. The intensity of third-harmonic signal can then be calculated using Eq. (10). This is plotted as a function of the energy of the second-harmonic pulse for two values of the pulse energies for the fundamental beam in Figure 6. The experimental data are in good agreements with the simulated values using the above theoretical formulism.

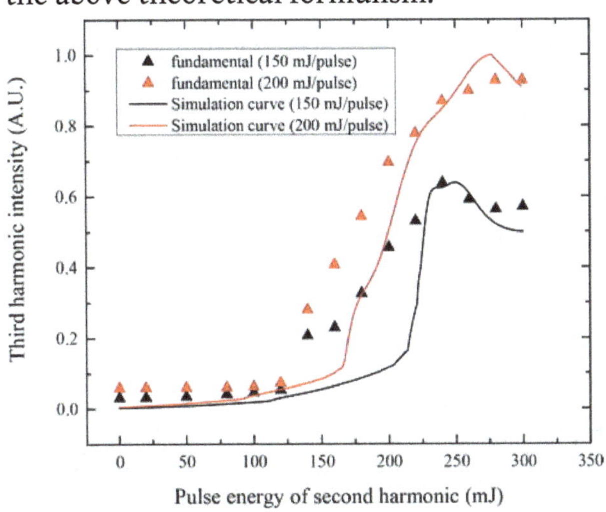

Figure 6. The comparison of the experimental and simulation results for two-color excited third-harmonic signal in argon as a function of the second-harmonic pulse energy.

The four-wave mixing process is dominant in the third-harmonic signal. For our experimental conditions, the THG component is approximately 10^{-4} that of the FWM process. In the low plasma density limit, the FWM term can be written as $I_{3II} \propto (\chi^{(3)}_{II,p})^2 \propto N_e^{\,2}$. This indicates that the enhancement of the third-harmonic signal is due to the plasma-enhanced susceptibility for the FWM process. On the other hand, when the plasma density is high enough, the wave-vector mismatch Δk becomes significant due to the plasma-induced refractive index change, which is linearly proportional to the plasma density.

Thus, in the high plasma density limit, the four-wave mixing term becomes $I_{3II} \propto \mathrm{sinc}^2\,(\Delta k_{213,p} L/2)$. This is one of the reasons why the third-harmonic signal reaches saturation at high plasma density. Consequently, the TH signal is enhanced when the pulse energy of the fundamental output is reduced. This is consistent with the experimental findings for fundamental pulse energies of 150 and 200 mJ. Nevertheless, the experimental data does not align with the theoretically predicted plasma enhancement threshold. The threshold's relationship to the step-like enhanced ionization probability may account for this. When the effective photon energy (effective frequency) reaches the threshold of this process, the step-like behavior caused by new absorption processes becomes dominant. Nevertheless, there are numerous quantum processes that involve the absorption of multiple photons at frequencies of ω and 2ω. The ionization rate varies among the various quantum processes. The ionization probability of the distinct quantum process is also altered when the power ratio of the two-color field is altered. It could be contended that the ionization rate fluctuates continuously with the second-harmonic pulse energy as the plasma density increases, rather than in a step-like manner. This should subsequently cause the threshold pulse energy to change.

4. LASER-MATERIAL PROCESSING WITH MULTI-COLOR SYNTHESIZED LIGHT FIELD

In recent years, there has been a growing trend toward the use of high-energy laser beams for the refining and fabrication of materials and devices. To name a few, these include the fabrication of micro electro mechanical systems, optoelectronic components, biomedical micro fluid chips, silicon chip processing, electronic packages, and circuit board machining. During laser processing of materials, two types of mechanisms are present: a photo-thermal and a photo-chemical mechanism. A thermal source is employed in the photo-thermal mechanism, which involves the focalization of laser beams with a high power density on an object for an extended period. Thermal conduction facilitates the transmission of energy from the object's surface to its interior. The deposited thermal energy subsequently causes a portion of the object to dissolve or vaporize. The laser point is relocated to a different region of the workpiece in preparation for additional processing. In the photochemical mechanism, the bonding of molecules in the material to be processed is disrupted following the absorption of one or more photons. This process causes electrons to transition between energy levels, resulting in the potential for molecular bonds to fail [12, 13].

The laser is selected in laser processing based on the material's melting point, thermal diffusion, and energy absorption. For instance, lasers with suitable wavelengths are employed to conduct ablation on a variety of materials. Consequently, it is intriguing to determine whether the synthesized waveforms that were proposed and demonstrated in our research could be advantageous for laser processing.

Recent reports have documented the ablation of materials using multiple lasers, such as lasers with dual colors [14, 15, 16, 17, 18]. Implementation of coherent or incoherent summation of multi-color beams is feasible. Théberge and Chin [14] observed an increase in the volume of the vaporized material as a result of the incoherent summation of two femtosecond and nanosecond class pulsed lasers. The ns pulses were employed to exploit the free electrons and defect states that were induced by intensive fs pulses in this work. Okoshi and Inoue [15] demonstrated in a separate study that the Ti: sapphire laser with a relative fluence ratio of 1/39 was capable of etching polyethylene (PE) much deeper and more rapidly when superimposed fs pulses at the fundamental (ω) and a small fraction of the second-harmonics (2ω) output were used. The observed phenomena were ascribed to the increased photon energy of 2ω pulses, which can disrupt the chemical bonds of PE and create a modified layer of PE on the ablated surface. Nevertheless, the article did not address the temporal dynamics of the laser ablation process. Conversely, the coherent summation of dual-color pulses at zero delay resulted in the enhancement of absorption/reflection in fused silica [16]. This is due to the formation of defect states or the generation of free electron ions in the material. The ablation process was reported for silicon in the context of nanosecond and picosecond laser pulses, where a small portion of the (2ω) beam has the potential to excite electrons into the conduction band [17]. This effect was rendered insignificant for femtosecond pulses due to the creation of a sufficient population in the conduction band by multiphoton absorption in silicon. Nevertheless, the aforementioned works all account for relatively lengthy time delays between the beams of two colors (approximately picoseconds) on the scale of carrier lifetime.

We observe that the physical mechanism of intense-field photoionization in the gas phase was investigated by implementing tunable relative-phase control between the two dual-color exciting lasers [19, 20, 21]. The relative phase of the dual-colors is indeed a determining factor in the number of photoelectrons generated in a regime where both multiphoton and tunneling ionization mechanisms are present, as reported by Schumacher and Bucksbaum [19]. Subsequently, Gao et al. [20] demonstrated that the observed phase-dependence is indicative of quantum interference (QI) between the various channels, which correspond to varying numbers of photons. The threshold of plasma creation in the material to be ablated has been considerably reduced with the use of ns infrared laser pulses and its second-harmonic counterpart, in comparison to monochromatic excitation, as recently identified [21]. The field-dependence of the ionization cross section was the explanation for the observed phenomenon. The ablation study of metals under phase-controlled dual color ns pulses with a relative delay between the two colors of less than one oscillation period is the primary focus of this work.

The multi-color laser system employed in this study yielded preliminary results in the drilling of copper and stainless steel (refer to Figure 7). The cascaded layout is used to organize the nonlinear optical crystals for harmonic generation. The crystals are classified as KD*P type II for the second harmonic, ω^2, KD*P type I for the third harmonic, ω^3, BBO type I for the fourth harmonic, ω^4, and BBO type I for the fifth harmonic, ω^5, respectively. Consequently, the five-color frequency components' spectra extend from the near infrared (1064 nm) to the ultraviolet (213 nm). The cascaded harmonic generation arrangement was implemented to guarantee that the second-order nonlinear optical process occurred in a collinear manner, resulting in the spatial overlap of the fundamental and harmonic components. Each harmonic's relative phase and amplitude can be independently adjusted.

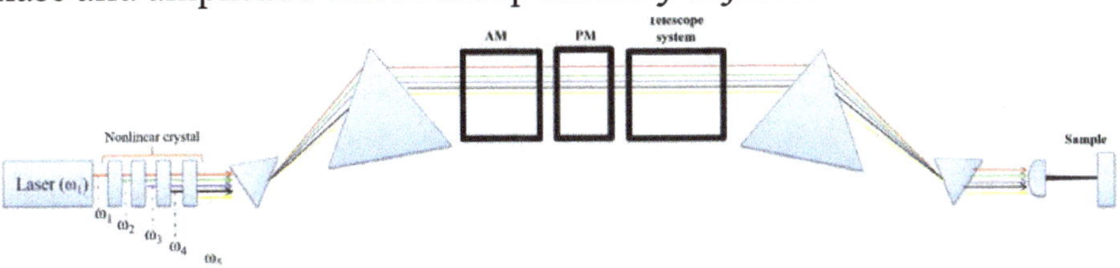

Figure 7. A schematic view showing a multi-color harmonic synthesized laser system for laser processing.

We studied two-color laser ablation of cooper and steel to demonstrate the feasibility of the approach. In the plane-wave approximation, the synthesized dual-color laser field can be written as,

$$E_{\text{Total}}(t) = E_{\omega}(t) \times \cos[\omega t] + E_{2\omega}(t) \times \cos[2\omega t + \theta], \tag{19}$$

where $E_{\omega}(t) = E_{0,\omega}\ exp[(-2\ln 2)t^2/\tau_{\omega}^{\ 2}]$ and $E_{2\omega}(t) = E_{0,2\omega}\ exp[(-2\ln 2)t^2/\tau_{2\omega}^{\ 2}]$ are the amplitudes of fundamental and second-harmonic optical fields with pulse durations, $\tau_{\omega,}$ and $\tau_{2\omega}$, both assumed to be around 15 ns; where ω, and θ are the angular frequency and relative phase between two fields, respectively. Because the ns pulse duration we can neglect a group velocity mismatch when dual wavelength waveform propagates through a media. Therefore, the first and second-harmonic pulses are fully overlapped and the joint pulse has the same pulse duration around 15 ns. Figure 8(a) and (b) illustrate the simulated synthesized electric-filed waveforms and the instantaneous intensities using Eq. (19) for the relative phases of θ = 0, π/2, π, and 3π/2, respectively, and $E_{2\omega}/E_{\omega} \approx 0.68$ (they are selected to match one of the experimental parameters corresponded to the intensity ratio $P_{2\omega}/P_{\omega} \approx 0.47$). It is clear that the waveform of combined ω and 2ω fields are very sensitive to the relative phase between the two. As shown in Figure 8(a), the synthesized waveform E_{Total} has a symmetric shape for θ = 0 and θ = π. It exhibits, however, an asymmetric profile with enhanced electric filed during one-half cycle of its oscillation for θ equal to π/2 and 3π/2. In comparison Figure 8(b), the instantaneous intensity of the two-color field for θ = 0 and π are quite similar to the cases where θ = π/2 and 3π/2 except for a shift in the time domain, like for. In other words, the period of the change of instantaneous intensity can be considered to be π.

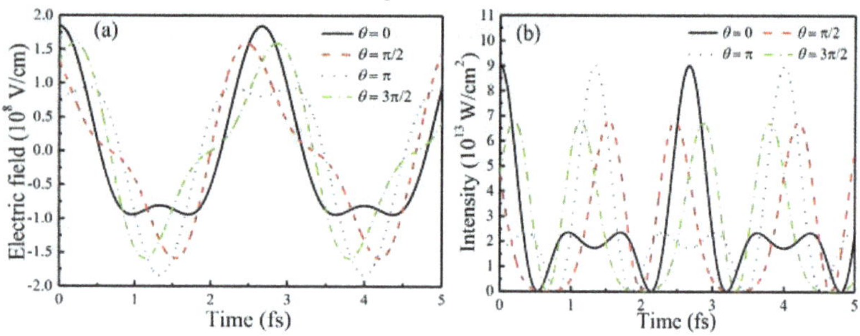

Figure 8. (a) Comparison of synthesized laser fields, and (b) instantaneous intensity of the fundamental wave (ω) and its second-harmonic (2ω) field with that of a ratio, E2ω/ Eω ≈ 0.68; shown are the cases of different relative phases θ = 0, π/2, π, and 3π/2, respectively.

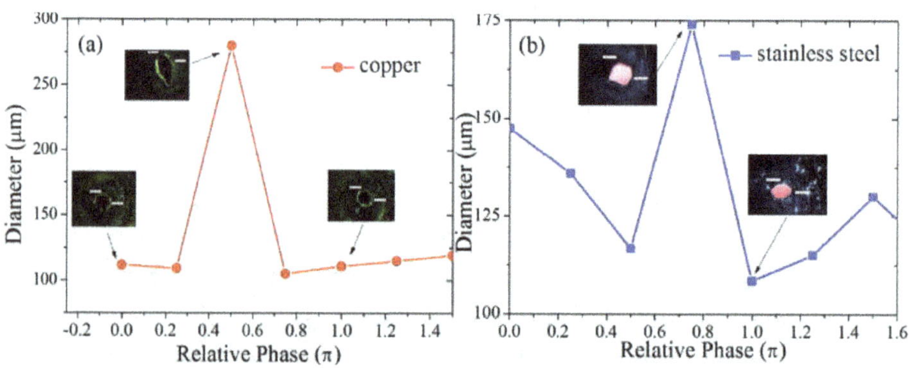

Figure 9. The diameters holes drilled in (a) copper, and (b) stainless steel by synthesized laser fields with different relative phases between the fundamental (ω_1) and second (ω_2) harmonics of the Q-switched laser. Pictures of drilled holes are also shown.

During the experiment, we applied the dual color ns pulses with the same total energy (~100 mJ) to 150 μm thick copper and stainless steel foil. We adjusted the phase modulators only, so, varied the relative phase between the harmonics. Then we measured the time required to make a pass through hole in a foil and estimated the ablation rate. In Figure 9(a), we have plotted diameters of holes drilled in copper sheetsas a function of relative phases between the fundamental (ω_1) and second (ω_2) harmonics of the synthesizing laser. Pictures of the drilled holes are also presented. Similar results for stainless steel are shown in Figure 9(b). Clearly, there is an optimal phase relationship between the two colors where higher instantaneous intensity causes higher ablation rate.

Figure 10(a)–(d) shows simulated results the peak strengths of the synthesized laser field with various relative phases ($\Delta\varphi$ = 0, 0.5π, π, and 1.5π) As can be seen in Figure 10(b) and (d), the synthesized laser field is expected to exhibit the higher peak strength at relative phases of $\Delta\varphi$ = 0.5π, and 1.5π. Therefore, ablation is expected to be more efficient for these waveforms.

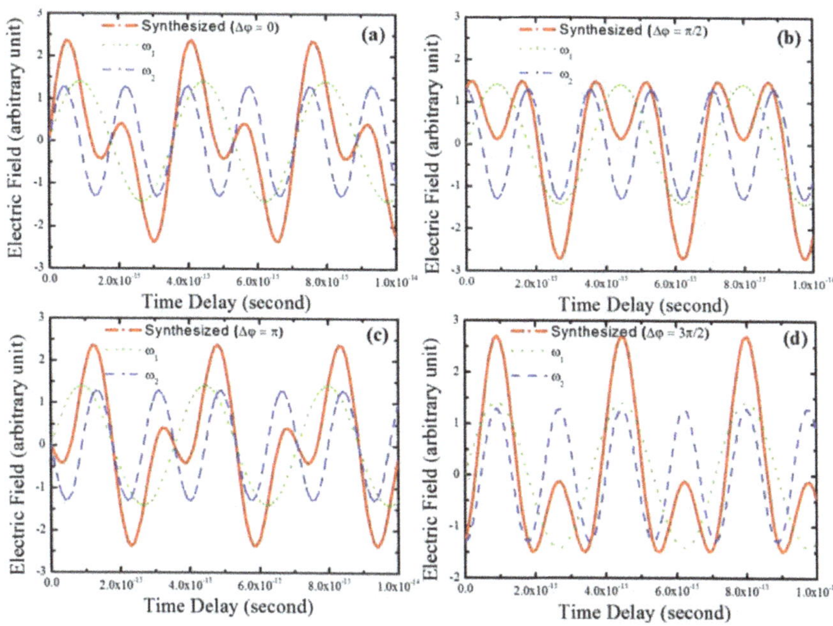

Figure 10. Numerical simulation of the peak strength of the laser waveform synthesized by two-color laser fields with various relative phases between the fundamental and the second harmonic (a) $\Delta\varphi$ = 0, (b) $\Delta\varphi$ = π/2, (c) $\Delta\varphi$ = π, and (d) $\Delta\varphi$ = 3π/2, respectively.

In another experiment, we fixed the exposure time at 10 s, varied the relative phase of the two-color beams and examined the ablated holes afterwards. In Figure 11(a), we have plotted diameters of holes drilled in copper sheets as a function of relative phases between the fundamental (ω_1) and second harmonics (ω_2) of the synthesizing laser. Pictures of the drilled holes are also presented. Similar results for stainless steel are shown in Figure 11(b). These data clearly show the dependence of ablation rate on the synthesized waveform, that is, relative phase of the fundamental (ω_1) and second harmonic (ω_2) of the single-frequency Nd:YAG laser.

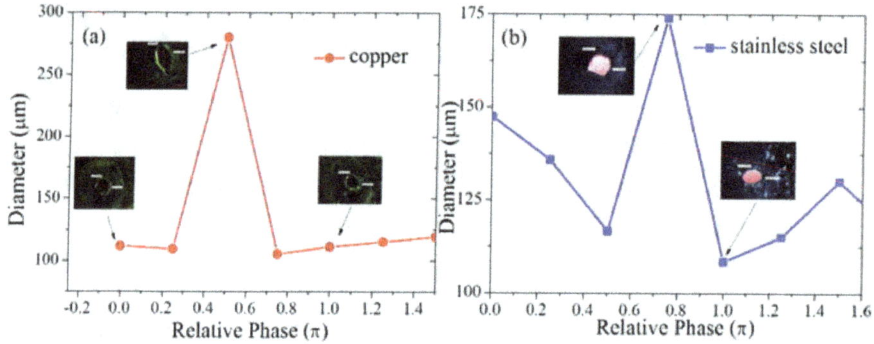

Figure 11. Diameters of holes drilled in (a) copper, and (b) stainless steel by synthesized laser fields with different relative phases between the fundamental (ω_1) and second (ω_2) harmonics of the Q-switched laser. Pictures of drilled holes are also shown.

5. CONCISE SUMMARY

We conducted a study on harmonic generation in inert gas systems using three-color waveform synthesis as an application of the high-power laser system built on synthesized waveforms. The interaction between a three-color beam and inert gases in third-order nonlinear optics can be employed to produce fourth to ninth harmonics of the laser fundamental output. For fourth-harmonic generation, there are three kinds of four-wave mixing processes: $\omega_4 = \omega_i + \omega_j + \omega_k$, $\omega_4 = \omega_i + \omega_j - \omega_k$, where i, j, k = 1, 2, 3. For fifth-harmonic generation, there are three possible processes: $\omega_5 = \omega_1 + \omega_2 + \omega_2$, $\omega_5 = \omega_1 + \omega_1 + \omega_3$, $\omega_5 = \omega_3 + \omega_3 - \omega_1$. There are two types of four-wave mixing procedures for the sixth and seventh harmonic, and so forth. The simulation results for fourth-harmonic generation using three-order nonlinear processes are presented in detail to serve as an illustration. The fourth-harmonic signal is demonstrated to fluctuate in accordance with the fundamental beam's phase.

In the past, we investigated the impact of the relative phases and intensities of the two-color pump on the third-order nonlinear frequency conversion process. The third-harmonic (TH) signal oscillates periodically with the relative phases of the two-color driving laser fields as a result of the interference of TH signals from a direct third-harmonic-generation (THG) channel and a four-wave mixing (FWM) channel. This is demonstrated. However, plasma can be produced through the ionization process in a high-intensity laser field. The plasma density in the multiphoton ionization region was estimated using the Perelomov, Popov, and Terent'ev (PPT) model, which considered the instantaneous laser field and laser frequency. We demonstrate that plasma plays a substantial role in the generation of the third-harmonic signal, based on the assumption that susceptibility and wave-vector mismatch are dependent on the plasma density. The experiments are in excellent agreement with the simulation results. Lastly, we presented preliminary data that suggests the synthesized two-color laser fields are capable of significantly improving the conversion efficiency of HHG and VUV spectra. Additionally, we demonstrated phase-sensitive two-color ablation of stainless steel and copper. Our findings demonstrate that the utilization of optimized waveforms enhances the efficiency of hole drilling.

AT A GLANCE

In the first half of this study, we describe the process of designing, constructing, and diagnosing a novel method for generating high-power attosecond pulses and arbitrary waveforms through multicolor synthesis. This chapter showcases many uses of this innovative source, including coherently regulated harmonic production and phase-sensitive two-color ablation of copper and stainless steel using the multicolor laser system.

REFERENCES

1. Pan C-L, Yang C-S, Zatazev A, Chen W-J, Lee C-K. Frequency-synthesized approach to high-power Attosecond pulse generation and applications (I). In: Harooni M, editor. High Power Laser Systems. Rijeka, Croatia: InTech; 2018. ISBN: 978-953-51-5267-5

2. Chen W-J, Wang H-Z, Lin R-Y, Lee C-K, Pan C-L. Attosecond pulse synthesis and arbitrary waveform generation with cascaded harmonics of an injection-seeded high-power Q-switched Nd:YAG laser. Laser Physics Letters. 2012;9(3):212

3. Suntsov S, Abdollahpour D, Papazoglou DG, Tzortzakis S. Efficient third-harmonic generation through tailored IR femtosecond laser pulse filamentation in air. Optics Express. 2009;17(5):3190-3195

4. Fedotov A, Gladkov S, Koroteev N, Zheltikov A. Highly efficient frequency tripling of laser radiation in a low-temperature laser-produced gaseous plasma. JOSA B. 1991;8(2):363-366

5. Feng L-B, Lu X, Xi T-T, Liu X-L, Li Y-T, Chen L-M, Ma J-L, Dong Q-L, Wang W-M, Sheng Z-M. Numerical studies of third-harmonic generation in laser filament in air perturbed by plasma spot. Physics of Plasmas. 2012;19(7):072305

6. Suntsov S, Abdollahpour D, Papazoglou D, Tzortzakis S. Filamentation-induced third-harmonic generation in air via plasma-enhanced third-order susceptibility. Physical Review A. 2010;81(3):033817

7. Rodríguez C, Sun Z,Wang Z, RudolphW. Characterization of laser-induced air plasmas by third harmonic generation. Optics Express. 2011;19(17):16115-16125

8. Mahon R, McIlrath TJ, Myerscough VP, Koopman DW. Third-Harmonic Generation in Argon, Krypton, and Xenon: Bandwidth Limitations in the Vicinity of Lyman-(alpha). Quantum Electronics, IEEE Journal. 1979;5(6):444-451

9. Marr G, West J. Absolute Photoionization Cross-Section Tables For Helium, Neon, Argon, and Krypton In The VUV Spectral Regions. Atomic Data and Nuclear Data Tables. 1976;18(5):497-508

10. Chang Z. Fundamentals of Attosecond Optics. 1st ed. CRC Press; February 16, 2011. ISBN-10: 1420089374. ISBN-13: 978-1420089370

11. Sapaev U, Husakou A, Herrmann J. Combined action of the bound-electron nonlinearity and the tunnel-ionization current in low-order harmonic generation in noble gases. Optics Express. 2013;21(21):25582-25591

12. Cain SR. A photothermal model for polymer ablation: Chemical modification. Journal of Physical Chemistry. 1993;97:051902-1-5

13. Pan C-L, Lin C-H, Yang C-S, Zaytsev A. Laser ablation of polymethylmethacrylate (PMMA) by phase-controlled femtosecond two-color synthesized waveforms. In: Yang D, editor. Applications of Laser Ablation—Thin Film Deposition, Nanomaterial Synthesis and Surface Modification. London, UK: InTech; 2016. ISBN: 978-953-51-2812-0, Print ISBN: 978-953-51-2811-3

14. Théberge F, Chin SL. Enhanced ablation of silica by the superposition of femtosecond and nanosecond laser pulses. Applied Physics A: Materials Science & Processing. 2005;80:1505

15. Okoshi M, Inoue N. Laser ablation of polymers using 395 nm and 790 nm femtosecond lasers. Applied Physics A: Materials Science & Processing. 2004;79:841

16. Chowdhury IH, Xu X, Weiner AM. Ultrafast two-color ablation of fused silica. Applied Physics A: Materials Science & Processing. 2006;83:49

17. Zoppel S, Merz R, Zehetner J, Reider GA. Enhancement of laser ablation yield by two color excitation. Applied Physics A: Materials Science & Processing. 2005;81:847

18. Zoppel S, Zehetner J, Reider GA. Two color laser ablation: Enhanced yield, improved machining. Applied Surface Science. 2007;253:7692

19. Schumacher DW, Bucksbaum PH. Phase dependence of intense-field ionization. Physical Review A. 1996;54(5):4271

20. Gao L, Li X, Fu P, Guo D-S. Phase-difference effect in two-color above-threshold ionization. Physical Review A. 1998;58(5):3807

21. Schwarz E, Reider GA. Laser-induced plasma by two-color excitation. Applied Physics B: Lasers and Optics. 2012;107(1):23

CHAPTER-9

FEMTOSECOND LASER ENERGY HARVESTING THEORY

1. INTRODUCTION: AN OVERVIEW

Over the past few decades, there has been a significant advancement in the ability to scale up the energy of femtosecond laser pulses. This development has made it possible to conduct high-energy physics experiments using tabletop setups in mid-level university laboratories [1, 2]. Consequently, the intensities of around 10^15 watts per square centimeter can be directly achieved using a mode-locked thin-disk laser oscillator that operates at a repetition rate above one megahertz [3, 4, 5, 6, 7, 8]. These systems are significantly simpler, more efficient, and cost-effective compared to traditional chirped-pulse amplifiers, specifically at the gigawatt pulse power level [9, 10]. In addition, a high repetition rate results in a signal/noise ratio improvement of 10^3 - 10^4 compared to a regular kilohertz chirped pulse amplifier. These oscillators are used in various applications such as gas ionization, high-harmonic generation, electron diffraction experiments, fabrication of fine structures in materials, characterization and control of electronic dynamics, metrology, ultra-sensitive spectroscopy, biophotonics, and biomedicine.

The concept of ultrashort pulse energy harvesting extends beyond the boundaries of laser-based applications and is relevant to the generation and control of larger coherent structures in a broader context [17]. High-energy ultrafast lasers are a valuable tool for investigating fundamental issues related to self-organization and nonlinear dynamics in systems that are far from thermodynamic equilibrium. These systems encompass a wide range of fields, including hydrodynamics, condensed matter physics, biology, and sociology [18, 19, 20]. This approach, which involves transferring complex dynamics to a simpler material context, can be referred to as "metaphoric" or "analog" modeling [21, 22]. It has been successful because of its high controllability, relative simplicity, and the unique potential for gathering statistics inherent in laser systems [1].

The concept of energy harvesting is founded on a fundamental equation: E = P av T res, where P av represents the average power in a resonator with an effective period T res. The scaling of ultrashort pulse energy can be determined by scaling the values of P av and/or T res, provided that a stable ultrashort pulse is generated spontaneously (referred to as mode-locking self-start condition) in a laser system. These parameters are complex and challenging for energy-scalable lasers, and they can significantly restrict the pulse energy and its width. This work will address two primary methods for energy harvesting at the femtosecond scale.

The first method relies on the distinctive ability of laser dissipative solitons (DS) [17, 23] to amass energy while maintaining stability [24, 25]. This presentation will provide fundamental methods for studying the energy-scaling laws of these systems, as well as provide an overview of the boundaries for energy and pulse width scalability. The second method relies on the storage of energy in an external high-Q resonator, known as an enhancement resonator (ER), which is simultaneously coupled with a femtosecond pulse oscillator [26, 27, 28]. Implementing this straightforward concept on a femtosecond scale encounters challenges because to the presence of nonlinear effects and group-delay dispersion (GDD), which disrupt the synchronization between a laser and ER. The following concerns will be delineated, and certain alterations to the ER technique will be suggested.

2. DS ENERGY SCALING

The DS energy and width scaling are connected closely with a duality between amplification of the *maximum number of laser modes* and simultaneous *spectral condensation*, i.e., the concentration of energy within the strongly confined spectral region. It is important that all excited and amplified modes must be phase-synchronized, i.e., coherent.

In a trivial model of laser, there exists a set of N - longitudinal resonator modes which are separated by the frequency interval of $\delta\omega = 2\pi / T_{res}$ and excited by a gain medium with the gain-bandwidth Ω : $N \sim \Omega / \delta\omega$ [29]. These modes are mutually phase-independent and incoherent, and a result of their interference $A = \Sigma_{n=-N/2}^{N/2} a_i \cos(\omega_0 + n\delta\omega t + n\delta\phi]$ is the irregular field beatings with the width of separated spikes $\sim 1 / N\delta\omega = 1 / \Omega$ (Figure 1(a); the inter-mode phase difference $\delta\phi$ is random). However, a fixed inter-mode phase difference results in regular spikes of the $\sim 1 / \Omega$ width with the peak power $\sim N^2$ and the repetition-period $= T_{res}$ (Figure 1(b)) [30]. The last phenomenon is called *mode-locking* (ML) and underlies a *coherent energy condensation* within short-time intervals. Respectively, the spectral width of each spike tends to $\Delta\Omega$.

However, this simple scheme faces many complications. Well, to be precise, a gain-band is not uniform (bell-shape like) and a mode, which is closest to a frequency ω_0 at gain maximum, has maximum amplification. Since laser gain is energy-saturable, this mode concentrates all energy and suppresses the competitive modes. This is a *mode selection* process. Therefore, a multimode generation leading to ultrashort pulse formation is not a genuine but *emergent* phenomenon which requires a multimode instability.

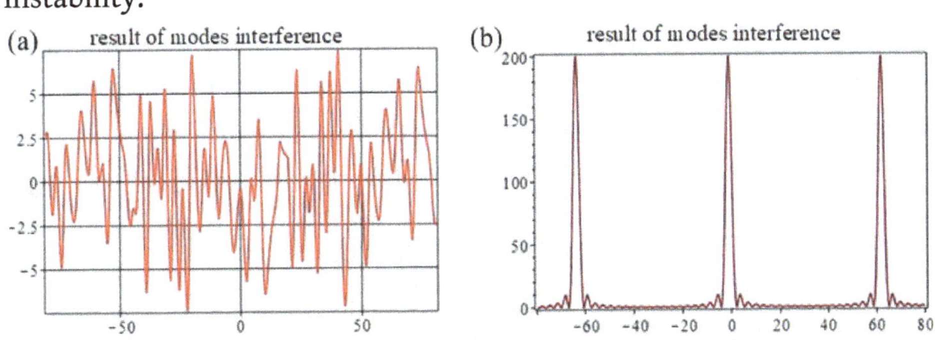

Figure 1. Interference of phase uncoupled (a) and locked (b) modes ($N = 20$) [30].

There are several possible mechanisms for such instability [31] which are closely connected with the issue of the ML *self-start*. Existing theories of the ML self-start predict a lot of effects involved in a laser pulse formation including mode-beatings [32, 33] and hole burning, induced refractive grating in an active medium [34], dynamic gain saturation [33, 35], parasitic reflections and absorption in a resonator [36], continuous-wave instability [37, 38] and Risken-Nummedal-Graham-Haken effect [39]. The *thermodynamic theory* of ML self-start has been developed, and it has been shown that the pulse appearance is a first-order phase transition, which is affected strongly by the laser noises distributed over a whole resonator period [40, 41, 42]. In any case, a stable ML requires whether a nonlinear resonant excitation by external periodic "force" (*active mode-locking*)[1] or a mutual mode coupling through an optical nonlinearity (*passive mode-locking*). The excitation of harmonics at $\pm n\delta\omega$ couples and synchronizes the adjacent modes and provides their phase-locking.

The mechanisms of ML are beyond the scopes of this work, and we will focus on the principles of the sustained ML energy-scalable regimes at femtosecond scale. The basic principle under consideration is to exploit DS [23] which is extremely stable in nonequilibrium dissipative environment [24, 25]. Since DS behaves like a soliton of integrable systems [17, 44], its dynamics can be described by some distributed nonlinear model. The most famous and studied one is based on the complex nonlinear Ginzburg-Landau equation which can be treated as a dissipative extension of the nonlinear Schrödinger equation [45].[2]

A very productive approach to the study of this class of equations is based on the so-called *variational approximation* (VA) [49, 50, 51]. The non-dissipative effects can be described by the Lagrangian density:

$$L = \frac{i}{2}\left(a^*(z,t)\partial_z a(z,t) - a(z,t)\partial_z a^*(z,t)\right) - \frac{1}{2}\gamma|a(z,t)|^4 + \frac{\beta_2}{2}\partial_t a(z,t)\partial_t a^*(z,t), \tag{1}$$

where a(z, t) is a complex slowly varying field amplitude, *t* and *z* are local time and propagation distance, respectively, γ is a self-phase modulation (SPM) coefficient and β_2 is a group-delay dispersion (GDD) coefficient. The Euler-Lagrange equation corresponding to Eq. (1) is the nonlinear Schrödinger equation:

$$i\partial_z a(z,t) = \frac{\beta_2}{2}\partial_{t,t} a(z,t) + \gamma|a(z,t)|^2 a(z,t). \tag{2}$$

Further, two different types of DS will be considered: (i) chirped-free

$$a(z,t) = \alpha(z)\operatorname{sech}(t/T(z))\exp\left(i\phi(z)\right), \tag{3}$$

and (ii) chirped pulses

$$a(z,t) = \alpha(z)\text{sech}(t/T(z))^{1+i\psi(z)}\exp\left(i\phi(z)\right), \tag{4}$$

where α, T, ψ and ϕ are DS amplitude, width, chirp, and phase-delay, respectively.

2.1. Chirped-free DS

VA in action looks like following. Substitution of the trial solution (3) into (1) with the subsequent integration over *t* results in the reduced Euler-Lagrange equations [52]:

$$4\gamma\alpha^2 + 6\partial_z\phi = \beta_2/T^2, \tag{5}$$

$$\beta_2/T^2 + 2\left(\gamma\alpha^2 + 3\partial_z\phi\right) = 0, \tag{6}$$

$$2\alpha^2\partial_z T + 4T\alpha\partial_z\alpha = 0. \tag{7}$$

Eq. (7) is the energy conservation law, and Eqs. (5) and (6) give the parameters of *Schrödinger soliton* if $\partial_z T = 0$, $\partial_z \alpha = 0$:

$$\alpha T = \sqrt{\beta_2/\gamma},\ q \equiv -\partial_z\phi = \frac{\beta_2}{2T^2}, \tag{8}$$

where the first expression is the soliton *area theorem,* and the last one defines the soliton *wave-number q*.

2.1.1. Perfectly saturable absorber

The effect of dissipative factors can be taken into account using the Ritz-Kantorovich method [51, 52, 53, 54] when the reduced Euler-Lagrange equations are driven by a dissipative "source" [52]:

$$\frac{d}{dz}\frac{\partial\int Ldt}{\partial(b_i)_z} - \frac{\partial\int Ldt}{\partial b_i} = 2\text{Re}\int Q\frac{\partial a^*}{\partial b_i}dt,$$
$$Q = -i\Gamma a + \frac{ip(a+\tau\partial_{t,t}a)}{1+\sigma\int|a|^2dt} + \frac{i\mu\zeta|a|^2a}{1+\zeta|a|^2}. \tag{9}$$

Here the reduced Lagrangian ∫Ldt is calculated using a trial function (i.e., Eq. (3)) with the parameters b i = α T ϕ . The dissipative factors are defined by a net-loss with the coefficient Γ, a small-signal gain ρ with the inverse saturation energy σ and a squared inverse gain bandwidth τ. The *self-amplitude modulation* (SAM) providing ML is described by an effective “perfectly saturable absorber” [24] with the modulation depth μ, and the inverse saturation power ζ.

A solution obtained from Eqs. (3) and (9) gives the expressions for area theorem and phase delay corresponding to the Schrödinger soliton. But the DS amplitude is defined by dissipative factors [52]:

$$\frac{\mu \log\left(\frac{1+\alpha^2-\alpha\sqrt{1+\alpha^2}}{1+\alpha^2+\alpha\sqrt{1+\alpha^2}}\right)}{\alpha\sqrt{1+\alpha^2}} + 2(\Sigma+\mu) - \frac{2}{3}C\alpha^2 = 0, \tag{10}$$

where $\Sigma = \frac{\rho}{1+2\alpha^2 T\sigma} - \Gamma, C = \frac{\rho\tau\gamma}{\beta_2\zeta}\left(1+2\alpha^2 T\sigma\right)^{-1}$. The peak power α^2 in Eq. (10) is normalized to ζ. It is convenient to assume that a laser operates in the vicinity of a threshold in steady-state regime: $\frac{\rho}{1+2\alpha^2 T\sigma} \approx \Gamma$ and $\Sigma \approx 0, C \approx \frac{\Gamma\tau\gamma}{\beta_2\zeta}$.

The marginal stability condition $\Sigma = 0$ defines a stability of DS against continuous-wave or multiple pulse generation [30, 55], and the DS approaching this stability threshold has a minimum width and a best asymptotical energy scalability $(E = 2\alpha^2 T\zeta/\sqrt{\tau\Gamma} \gg 1)$ [52]:

$$\begin{aligned} &C \to \text{const} \times \sqrt{\mu\tau\Gamma}/E\zeta,\ E \to \text{const} \times \frac{\beta_2}{\gamma}\sqrt{\frac{\mu}{\tau\Gamma}}, \\ &\alpha^2 \to \left(\frac{\text{const}}{2}\right)^2 \frac{\mu\beta_2}{\tau\Gamma\gamma},\ T \to \frac{2}{\text{const}}\sqrt{\frac{\tau\Gamma}{\mu}},\ \text{const} \approx 3.535. \end{aligned} \tag{11}$$

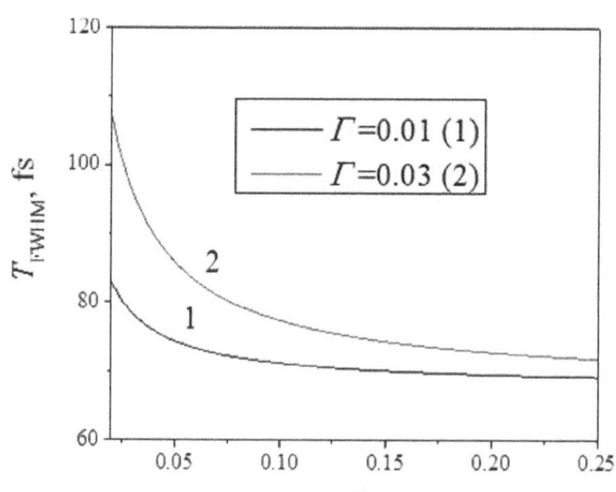

Figure 2. **Dependence of the asymptotic full width at half maximum (FWHM) T_{FWHM} on the modulation depth μ for different net-loss coefficients Γ. The gain bandwidth of 5.3 THz corresponds to a Yb: YAG thin-disk active medium.**

These scaling laws demonstrate main principles of chirped-free pulse energy harvesting. Of course, these dependencies can be considered as only qualitative ones. Nevertheless, they demonstrate that the asymptotic DS energy scales $\propto \beta_2$, and the minimum pulse width is defined by not only the medium gain bandwidth $\propto 1/\tau$ but the net-loss Γ and the modulation depth μ, as well (Figure 2).

Thus, the pulse can be squeezed by scaling of modulation depth with a parallel decrease of the stabilizing dispersion. Additional pulse shortening can be provided by net-loss lowering (see Figure 2). These tendencies are quite reasonable because the selective spectral properties of an active medium are defined by not the gain for a small signal but by the saturated gain $\approx \Gamma$ near the pulse stability threshold (i.e., since no gain, no gain induced spectral selection). Simultaneously, the modulation depth defines an inter-mode coupling strength that favors ML and, thereby, pulse spectrum broadening with μ-growth.

Eqs. (8) and (10) demonstrate that an approach to the threshold C ($\Sigma \to 0$) as well as a higher $E \propto \zeta/\sqrt{\tau}$ minimize the pulse width T (Figure 3).

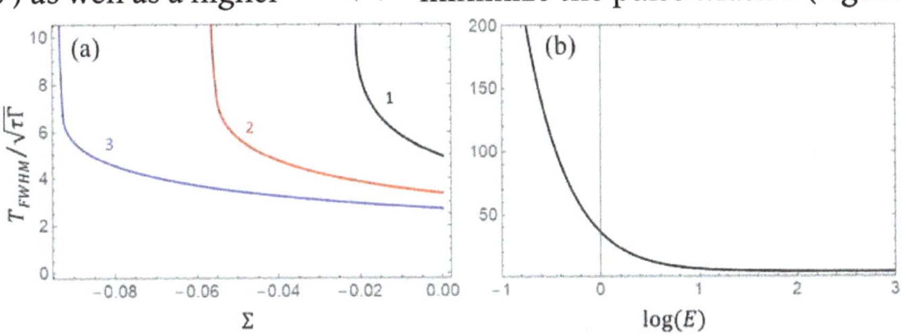

Figure 3. Dependence of the normalized T_{FWHM} on Σ (a) for μ = 0.05 (1), 0.1 (2), 0.15 (3), C = 0.01 ; and on the normalized energy E (b) for μ = 0.05, Σ = 0.

Properties of DS are described by the so-called *master diagram* [24] which represents the DS parametric space and is shown in Figure 4(a). There are two DS-solutions of Eqs. (1), (3), (9) and (10): (i) upper branch (i.e., the branch with the larger C for fixed E and Σ, see Figure 4(a)) corresponds to the above considered *energy-scalable* DS. The energy scalability for this type of soliton is accompanied by minimization of its width (Figures 3 and 4(b); i.e., $\lim_{E\to\infty,\,\Sigma\to 0} T = \frac{2}{\text{const}}\sqrt{\frac{\tau\Gamma}{\mu}}$, see Eq. (11)) and, respectively, by the growth of peak power. Namely, this branch has a threshold of marginal stability Σ = 0 (Figure 4(a); curve 1). (ii) lower branch (Figure 4(a)) corresponds to a DS energy scalability provided by its width growth (Figure 4(b); $\lim_{E\to\infty,\,\Sigma<0} T = \infty$). Thus, this DS branch is unpractical for energy scaling because the broad chirp-free pulse would require an additional nonlinear mechanism for external compression. One can name this branch as *energy-unscalable*.

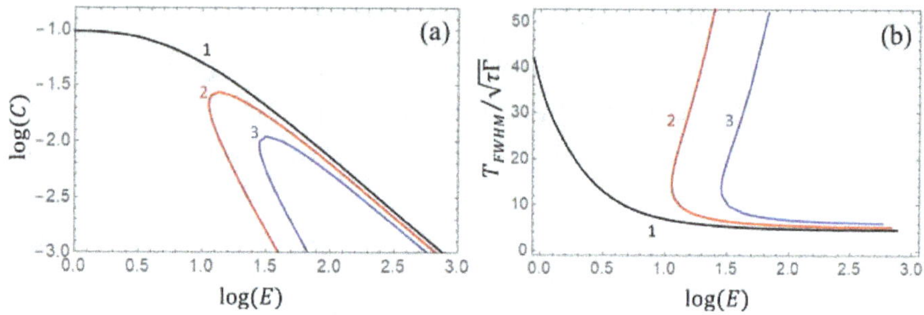

Figure 4. Master diagram (a) and the corresponding DS widths (b). Σ = 0 (1), -0.01 (2), and -0.02 (3), μ = 0.05.

A fundamental property of the DS solutions presented is their stability. The Vakhitov-Kolokolov stability criterion $dE / dq > 0$ [56, 57] demonstrates the stability of both branches of DS (see Eq. (8)):

$$\frac{dE}{dq} = \frac{d}{dq}\left(2\alpha^2 T\right) = \frac{d}{dq}\left(2\sqrt{2\beta_2 q}/\gamma\right) = \frac{\sqrt{2\beta_2/q}}{\gamma} > 0. \tag{12}$$

It means physically that an energy scalability of DS does not suffer from soliton collapse and both DS branches are feasible.

As was mentioned above, the net-gain Σ is energy-dependent, and such a dependence has to be taken into account. In the neighborhood of the laser threshold where $\Sigma\, E \approx 0$, one may expand the net-gain coefficient near a threshold energy $\Xi = \sigma^{-1} (\rho / \Gamma - 1)$:

$$\Sigma = \frac{d}{dE}\left(\frac{\rho}{1+\sigma E} - \Gamma\right)\bigg|_{E=\Xi} + O\left(\frac{d^2}{dE^2}\right) \approx \delta\left(\frac{E}{\Xi} - 1\right), \tag{13}$$

where $\delta = -\Gamma^2 \Xi \sigma / \rho$

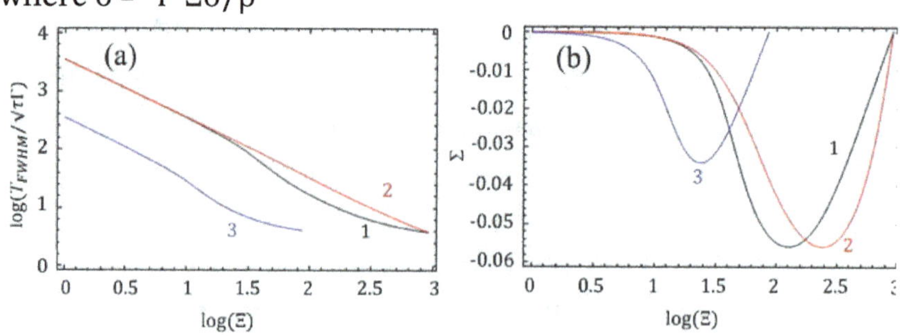

Figure 5. The dimensionless DS widths (a) and the corresponding net-gain (b) in dependence on the threshold energy Ξ for δ = -0.05, C = 10^{-3} (1); δ = -0.5, C = 10^{-3} (2); and δ = -0.05, C = 10^{-2} (3). μ = 0.07.

Figure 5(a) shows the DS dependence on the threshold energy for a fixed control parameter *C* in the presence of gain saturation. DS squeezes with energy, and such a squeezing is confined by the stability criterion $\Sigma < 0$. Simultaneously, Σ decreases from 0 with energy (Figure 5(b)), that corresponds to the "energy unscalable" DS, with the subsequent growth up to 0, that corresponds to the "energy scalable" DS. Thus, there is not a "switch" between two different types of DS in a real-world laser system which behaves quite smoothly with energy.

2.1.2. Cubic-quintic SAM

Physically, this type of SAM describes approximately an action of nonlinear polarization rotation, which is a typical ML mechanism for fiber lasers, or a so-called "soft aperture" Kerr-lens ML typical for solid-state lasers [24, 25]. In this case, loss saturation switches to the loss growth at $\alpha^2 = 1/2\zeta$ and

$$Q = -i\Gamma a + \frac{ip(a + \tau\partial_{t,t}a)}{1 + \sigma\int |a|^2 dt} + i\kappa\left(1 - \zeta|a|^2\right)|a|^2 a. \quad (14)$$

The κ-parameter plays a role of the inverse loss saturation power, and the modulation depth is $\mu = \kappa/4\zeta$.

The solutions for *T* and *q* correspond to Eq. (8), but the two-branch solution for DS peak power can be expressed in an explicit form:

$$\alpha^2 = \frac{5}{16}\left(2 - C \mp \sqrt{(2 - C)^2 + 96\zeta\Sigma/5\kappa}\right) \quad (15)$$

(power, time and energy are normalized to ζ, $\kappa/\zeta\tau\Gamma$ and $\kappa\zeta/\tau\Gamma$, respectively, and $C = \tau\Gamma\gamma/\beta 2\kappa$). These branches are separated by the energy curve E = 5 2 - C / C / 2 (curve 1 in Figure 6), and are shown in Figure 6.

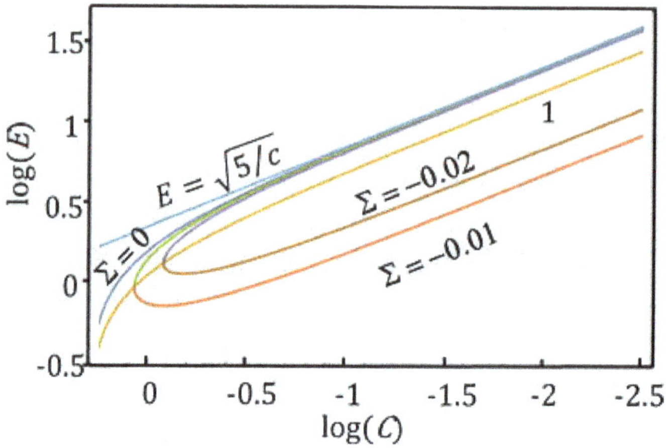

Figure 6. Master diagram for the chirp-free DS and the cubic-quintic SAM. Curve 1 divides two different branches of DS. The upper line E = 5 / C corresponds to asymptotical energy scaling law.

The asymptotic scaling laws for this type of SAM are:

$$C \to \frac{5\tau\Gamma}{4\mu\zeta^2 E^2}, E \to \sqrt{\frac{20\mu\beta_2}{\kappa\gamma}} = \sqrt{\frac{5\beta_2}{\zeta\gamma}},$$
$$\alpha^2 \to \frac{5\mu}{\kappa}, T \to \sqrt{\frac{\kappa\beta_2}{5\gamma\mu}}. \tag{16}$$

One can see that the energy scaling is provided by the DS width (not power) scaling that is a natural consequence of the peak power confinement α 2 = 1 / 2 ζ imposed by a SAM saturation[3]. Both branches behave quite congruently in this case (Figure 6).

2.1.3. Energy harvesting of the chirp-free DS at femtosecond scale

Most promising devices realizing the femtosecond-pulse energy scalability are thin-disk solid-state lasers [5, 6, 24, 58] which provide an excellent average power scaling and controllable nonlinear effects limiting the energy scalability in fiber oscillators [25]. Nevertheless, there are some main obstacles for further energy harvesting at femtosecond scale for such a type of devices. (i) a traditional ML mechanism uses the structured semiconductor devices (so-called *semiconductor saturable absorber mirrors, SESAM*) which (α) have a slow (~100 fs) response time; (β) a complicated and hardly-controllable kinematics including higher-order nonlinear effects, non-saturable losses, temperature and radiational damage, etc.; (γ) SAM and SPM effects are decoupled for this type of ML that requires growth of GDD for DS stabilization in accordance with the area theorem (Eq. (8)). (ii) As was shown (Eq. (11)), the minimum pulse width is $\propto \sqrt{\tau\Gamma/\mu}$ so that using the media with the broader gain band would provide a pulse shortening down to sub-100 fs [59]. However, such media with good optical quality are not widely available and technologically advanced.

Nevertheless, an alternative approach to energy scalability of femtosecond pulses has been demonstrated in [7]. (i) ML mechanism can be provided by an instantaneous self-focusing (Kerr-lensing) induced by a set of nonlinear crystals inside a laser resonator. (ii) Such a mechanism combines both SAM and SPM that enhances the SAM parameters (μ and ζ) in parallel with the SPM one (γ, see Eq. (11)). As a result, the GDD value can be reduced in parallel with the DS shortening in agreement with the area theorem. (iii) A real-world gain band profile is *Lorentzian*, not *Gaussian* as in Eqs. (9) and (14).

The Lorentzian gain profile can be taken into account by using the numerical simulations of the generalized complex nonlinear Ginzburg-Landau equation [60, 61]:

$$\partial_z a = -\Gamma a + \frac{\mu\zeta|a|^2}{1+\zeta|a|^2}a - i\left(\frac{\beta_2}{2}\partial_{t,t}a + \gamma|a|^2 a\right) + \frac{\rho\Omega}{2\left(1+\sigma\int|a|^2 dt\right)}\int_{-\infty}^{t} a\left(z,t'\right)\exp\left[-\Omega\left(t-t'\right)\right]dt', \quad (17)$$

where a characteristic gain bandwidth is $\Omega \propto 1/\tau$.

As was shown in [60], the Lorentzian gain profile gives more efficient amplification and broader spectrum than the Gaussian one. Additionally, an inherent gain dispersion shifts the DS spectrum and affects its shape [62]. The numerically obtained pulse spectra for different modulation depths are shown in Figure 7. One can see a pronounced spectrum broadening and, correspondingly, pulse shortening with the modulation depth growth.

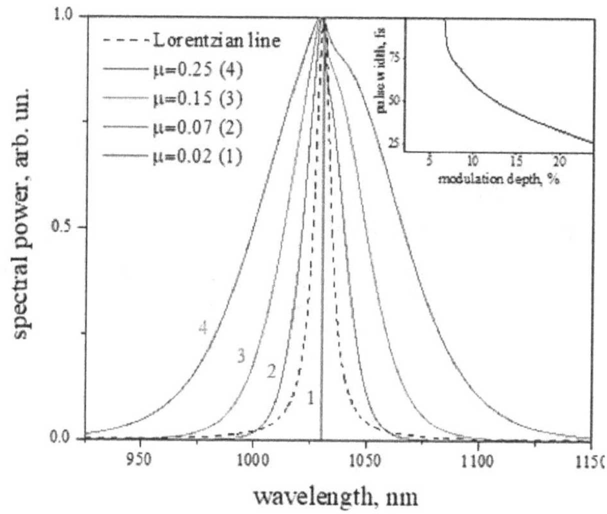

Figure 7. Numerical spectral profiles for different modulation depths μ (inset shows the dependence of T_{FWHM}). GDD corresponds to the stability threshold, $\zeta = \gamma = 1.35\ MW^{-1}$, $\delta = -0.05$. The Lorentzian gain band of 5.3 THz (dashed line) corresponds to a Yb: YAG, the output energy E out is of ≈ 0.011 μJ for 3% output coupler.

The dependences of minimum DS width and corresponding stabilizing GDD on the modulation depth for different values of the SAM saturation power ζ are shown in Figure 8.

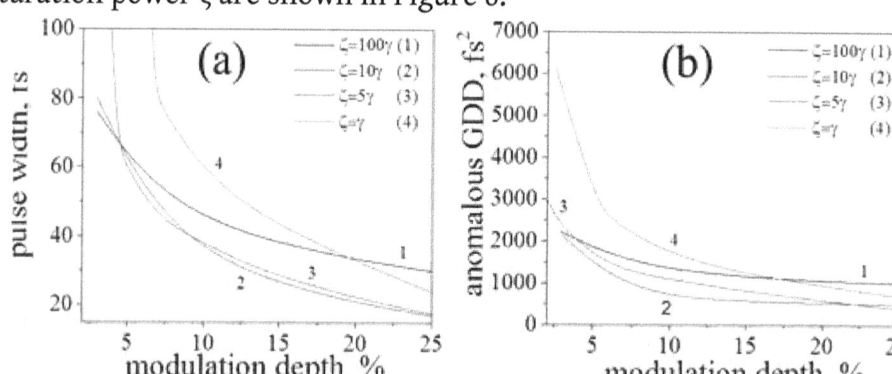

Figure 8. Pulse width T FWHM vs. modulation depth (a) along the boundary GDD (b) in dependence on the saturation parameter ζ . E out ≈ 0.011 – 0.014 μJ and other parameters correspond to Figure 7.

One can see that the DS shortens with μ in qualitative agreement with the analytical results presented above. Simultaneously, the DS spectrum is noticeably *broader* than the gain band that provides a generation of sub-50 fs pulses at the MW peak power level directly from an oscillator. Since a pulse is chirp-free, the threshold stabilizing anomalous GDD decreases, as well, in agreement with the soliton area theorem. Simultaneously, there is the nonmonotonic dependence of DS width on the SAM parameter ζ so that T decreases initially and then increases with the ζ-decrease, i.e., with the saturation power growth. The growth of saturation power (ζ-decrease) causes a threshold-like increase of pulse width and stabilizing GDD for small modulation depths μ.

Figure 9 illustrates the energy-dependence of the DS width and the stabilizing GDD for a low net-loss. The short pulses are possible in this case, as well, but a small modulation depth does not allow a substantial DS shortening.

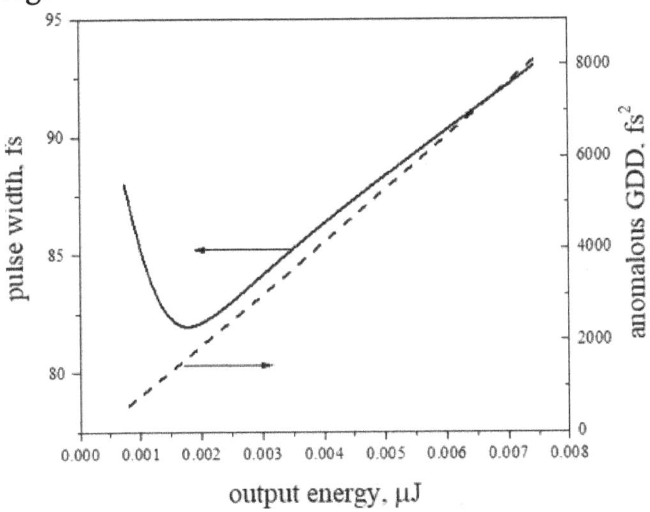

Figure 9. A low-loss regime with $\delta = -0.05/3$, $\zeta = 100\gamma$, 1% output coupler, $\mu = 2\%$.

2.2. Chirped DS

Chirped DS demonstrates a high potential for energy harvesting in both solid-state and fiber lasers [24, 25] due to enhanced stability provided by well-structured energy redistribution inside a pulse. An energy scalability results from the DS stretching that limits its peak power and, thereby, suppresses an instability caused by nonlinearity. This factor is especially important for all-fiber lasers, where the strong contribution of nonlinear effects is inevitable with T res -growth. VA predicts the following energy-scaling laws[4]:

$$E \propto |\beta_2|/C\sqrt{\tau\Gamma},\ E \propto |\beta_2|/\sqrt{\tau\Gamma} \tag{18}$$

for the SAM described by Eqs. (9) and (14), respectively [24, 25].

The chirped DS accumulates energy $\propto \psi$ that allows using a so-called *adiabatic theory* for $\psi \gg 1$ [24, 63, 64] which predicts a perfect energy scalability or a *DS resonance* [65] for the cubic-quintic SAM (Eq. (14)). That means that energy can be scaled infinitely for $C = 1/3$ due to pulse stretching and simultaneous *spectral condensation*:

$$\lim_{C \to 1/3} \begin{cases} E \to \infty \\ \alpha^2 \to 1/\zeta \\ \Delta \to \sqrt{2\gamma/\beta\zeta} \\ \Theta \to 0 \ (\text{"spectral condensation"}) \end{cases}, \tag{19}$$

where the DS spectral profile is described by a truncated Lorentzian function:

$$p(\omega) = \frac{6\pi\gamma}{\kappa\zeta}\frac{H(\Delta^2 - \omega^2)}{\omega^2 + \Theta^2} \tag{20}$$

(here H is the Heaviside's function).

For the SAM presented by Eq. (9), the adiabatic theory predicts the energy-scaling law in the form of [66][5]:

$$E \propto \frac{|\beta_2|}{\zeta\sqrt{\tau}}\left[\frac{3}{C}\sqrt{1-C} + \operatorname{arctanh}\left(\sqrt{1-C}\right)\right] \tag{21}$$

that gives the first expression in Eq. (18) in the $C \to 0$ (i.e., $E \to \infty$) limit.

Since the chirped DS energy scaling is provided by its stretching $\propto \psi$, this process is reversible so that an output DS can be compressed by a factor $\approx 1/\psi$. Nevertheless, some energy loss occurs with such compression due to *nonuniformity* of DS chirp [68][6] that requires optimizing the DS and pulse compressor parameters.

2.3. Main obstacles to the DS energy harvesting

The DS energy harvesting in both chirped and chirp-free regimes has a common problem of *ML self-starting*. The DS stability is a necessary but not sufficient condition of its existence because it must develop from some stochastic process in a laser (eventually, from a quantum noise). Existing theories of the ML self-start [32, 33, 34, 35, 36, 37, 38, 39, 40, 41, 42] predict that a lot of effects are involved in a pulse formation. However, a spontaneous formation of the DS from noise (the DS self-start) as a general problem has not been studied in depth. In optics, this is often considered as a technical issue, because here one can use one of the proven ML techniques to guarantee self-starting. After the initial kick, however, the DS evolves by itself, and recent experiments have shown controversial results: in high-power solid-state lasers, the strong oscillations (*Q-switching*) during the DS buildup dynamics hinders the DS self-start [69], while in a fiber laser, such oscillations can accelerate the self-start [70]. That is obviously connected to co-existence of nonlinearities with different time scales: instantaneous non-dissipative SPM, and non-instantaneous dissipative nonlinearities like stimulated Raman scattering (SRS), saturable absorber losses, and gain saturation. This issue is especially intriguing, as the dynamic gain saturation can provide a supplementary mechanism of DS formation [71].

The growing nonlinearity results in quite nontrivial modification of dynamics [25] and causes whether DS stabilization or its chaotization [72, 73, 74]. For example, the practically relevant Yb-based thin-disk lasers possess reduced instantaneous nonlinearity and longer gain relaxation times as compared to a bulk Ti: sapphire laser. In the latter case, the enhanced dynamic gain saturation can destabilize a much-desired high-energy DS [75, 76, 77]. As another example, the experiments demonstrated, that DS energy scaling in all-fiber fiber lasers is limited by energy loss due to SRS [78]. Nevertheless, SRS could play a positive role providing the generation of *dissipative Raman soliton* and suppressing the *optical turbulence* [79, 80, 81]. The connection of this phenomenon to the general issues of the turbulence theory waits for its exploration [82].

3. FEMTOSECOND PULSE ENHANCEMENT IN AN EXTERNAL RESONATOR

Using a high-finesse ER coupled with a mode-locked femtosecond laser is the method of energy storing allowing broadband absorption spectroscopy [83], high-harmonic generation and frequency comb generation up to the extreme ultraviolet frequency [84, 85].

Equations describing a coupling with ER are [86, 87]:

$$b_r = rb_{in} + \theta a_{in}, \quad a_r = \theta b_{in} - ra_{in}, \tag{22}$$

where b_r and b_{in} are the reflected and incident fields on the coupler from a side of femtosecond oscillator; a_r and a_{in} are the corresponding fields from a side of ER; and $\theta = \sqrt{1-r^2}$, r are the transmittance and reflection coefficients of a coupler, respectively. The field inside ER evolves as:

$$a_{in}(t) = \exp\left[-\Gamma + i\phi - \beta_1\frac{\partial}{\partial t} + i\sum_{m=2}^{N} i^m \beta_m \frac{\partial^m}{\partial t^m} + i\gamma|a_r(t)|^2\right]a_r(t), \tag{23}$$

where Γ is a net-loss coefficient; ϕ, β_1 are the phase and group-velocity delays, respectively; β_m are the mth-order GDD coefficients, and γ is the SPM coefficient.

In the absence of group-delay, GDD and SPM in ER, the energy, and power enhancement factors (Q_e and Q_p, respectively) are [87]:

$$Q_e = Q_p = \left(\frac{\theta}{1 - r\exp[-\Gamma]}\right)^2. \tag{24}$$

Under the condition of weak changes of the field during single cavity round-trip, Eqs. (22) and (23) can be rewritten in the form of the generalized driven nonlinear Schrödinger equation for the intracavity field $a(t)$ [88]:

$$\frac{\partial a(z,t)}{\partial z} = \left[-\Pi + i\Delta\phi - \beta_1\frac{\partial}{\partial t} + i\sum_{m=2} i^m \beta_m \frac{\partial^m}{\partial t^m} + i\gamma|a|^2\right]a + \theta\Phi(t), \tag{25}$$

where z is a cavity round-trip number, $\Pi = 1 - r + r\Gamma$, $\Phi(t)$ is an incident field amplitude and $\Delta\phi$ is a phase offset from the resonance $\phi = \pi$. In the absence of SPM (vacuum ER) but with GDD induced by resonator mirrors, the solutions for the energy and power enhancement factors are[7]:

$$Q_e = \frac{1}{2\pi\int_{-\infty}^{\infty}|\Phi(t)|^2 dt}\int_{-\infty}^{\infty}\left|\frac{\int_{-\infty}^{\infty}\Phi(t)e^{i\omega t}}{\Pi + i(\beta_1\omega - \beta_2\omega^2 + \beta_3\omega^3 - \beta_4\omega^4)}\right|^2 d\omega,$$
$$Q_p = \max\left\{\left|\frac{1}{2\pi}\int_{-\infty}^{\infty}\left[\frac{\int_{-\infty}^{\infty}\Phi(t)e^{i\omega t}dt}{\Pi + i(\beta_1\omega - \beta_2\omega^2 + \beta_3\omega^3 - \beta_4\omega^4)}\right]e^{-i\omega t}d\omega\right|^2\right\}. \tag{26}$$

Eq. (26) demonstrates reducing the enhancement factors due to GDD [87]. Indeed, β m ≠ 0 reduces the spectral power at the spectrum edges $\propto 1/(1+\omega^{2m}\beta_m^2)$ that worsens the spectral overlap between the pulses from a laser and ER. In combination with a chirp appearance and a pulse broadening in ER, these factors drop both Q_e and Q_p (Figure 10).

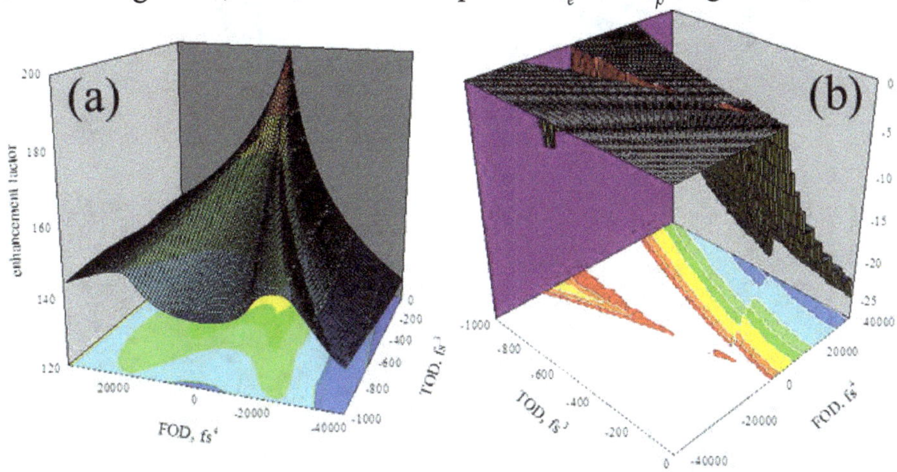

Figure 10. (a) Maximum Q_p inside ER in the dependence of the third-order dispersion (TOD, m = 3) and the fourth-order dispersion (FOD, m = 4) for $\gamma = 0$, $\Delta\phi = 0$, $\beta_1 \sim -0.5$ fs, Γ = 0.5% and 25 fs sech-shaped incident pulse at 790 nm central wavelength. (b) The corresponding optimal second-order dispersion (GDD, m = 2) [88].

It is clear that destructive action of higher-order GDD ($m > 2$ in Eq. (25)) on enhancement factor of ER grows with the pulse shortening so that a thorough dispersion-engineering of ER mirrors within a sufficiently broad spectral range is required [28, 87, 88].

Additionally, the enhancement factor control can be provided by realizing a soliton-like regime in the *nonlinear* ER with $\gamma \neq 0$ (Figure 11) [87, 88]. In the absence of higher-order dispersions (i.e., $m = 2$), Eq. (25) has an exact soliton-like solution $a(t) = \sqrt{P(t)}\exp(iqz)$:

$$P(t) = \left(\frac{\theta}{\Pi}\right)^2 \operatorname{sech}\left(\frac{t}{T}\right)^2, \beta_2 = -\left(\frac{\theta T}{\Pi}\right)^2 \frac{\gamma P(0)}{2}, \Delta\phi = -\left(\frac{\theta}{\Pi}\right)^2 \frac{\gamma P(0)}{2}, q = 0. \quad (27)$$

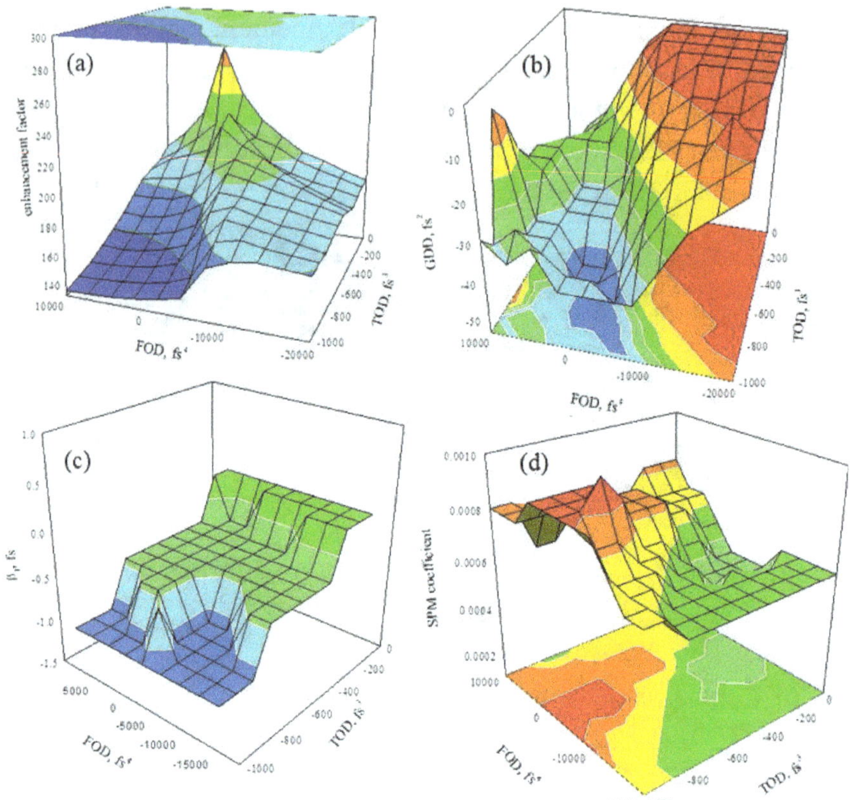

Figure 11. Dependence of the maximum Q_p on TOD and FOD (a) optimized by control of β_1 (c), β_2 (b), and the dimensionless SPM coefficient γP_{max} (d). Incident pulse T_{FWHM} is 25 fs, Γ = 0.25%.

[8]This soliton-like pulse can be perturbed by higher-order dispersions (m > 2) induced by the broad-band ER mirrors so that optimization of ER parameters is required in this case, as well (Figure 11).

A promising possibility of the Q_p-increase results from a *loss compensation* by a *gain* inside ER. In this case, Eq. (25) has to be supplemented by the term rρτ ∂ t , t a with the modified $\Pi = 1 - r + r \Gamma - \rho^9$ (see Eqs. (9) and (25)). One has to note, that ER remains *below* lasing threshold and a resonator mode in an active crystal (Ti: sapphire in our case) has to be sufficiently broad to suppress gain saturation and SPM. The soliton-like regime increases the enhancement factor and reduces the sensitivity to higher-order dispersions in this case, as well (Figure 12).

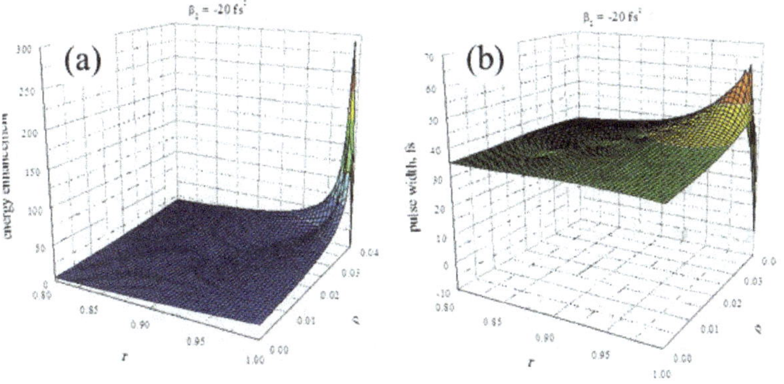

Figure 12. Q_e (a) and the pulse width (b) in the presence of gain, SPM, and GDD in ER. Γ = 0.04, the width of the incident 100 nJ pulse is of 30 fs, the laser beam size in the 2 mm Ti: sapphire active crystal is 1.1 mm.

4. CONCISE SUMMARY

This study provides a comprehensive examination and theoretical analysis of energy harvesting techniques at the femtosecond scale. The focus is on solid-state and fiber ML oscillators, with a special emphasis on the nonlinear ER. The fundamental idea is the existence of a dissipative soliton that enables precise manipulation of laser parameters to achieve controlled scaling of both extra-energy and spectral width. This method is highly productive for various applications and introduces high-energy physics into a "physics laboratory" where extreme parameters are achieved through careful adjustment rather than forceful manipulation. This tuning necessitates the use of interdisciplinary methods that enable the collection of power and energy at multiple scales. These methods are applicable in various fields, such as fundamental quantum mechanics, neuroscience, and sociology. They particularly involve the "quantum engineering" of Bose-Einstein and quasi-particle condensates. An further perspective involves studying the nonlinear dynamics of complex systems that are far from thermodynamic equilibrium. This study is centered on creating metaphorical models of these systems using simpler and more controlled laser systems.

AT A GLANCE

The energy scaling of femtosecond laser pulses has numerous applications in nanoscale micromachining, precision time-resolution spectroscopy, high-harmonic generation, surgery, and other fields. In addition to applied sciences and technology, there are basic uses of energy harvesting at the femtosecond scale. Specifically, it is feasible to investigate and manipulate the internal dynamics of atoms and molecules at the attosecond scale, while also accurately mapping quantum events with exceptional spatial and temporal precision. This combination of classical and quantum processes at the mesoscopic level offers fresh perspectives on fundamental aspects of quantum mechanics in open systems, with potential applications in the field of quantum computing. This study focuses on the investigation of femtosecond pulse energy extraction through the creation of dissipative solitons in solid-state and fiber mode-locked lasers, as well as the amplification of femtosecond pulses in an external resonator. The laws governing the scaling of femtosecond pulse energy, breadth, and spectrum are described in a clear and scientifically significant manner.

REFERENCES

1. Südmeyer T, Marchese SV, Hashimoto S, Baer CRE, Gingras G, Witzel B, Keller U. Femtosecond laser oscillators for high-field science. Nature Photonics. 2008;2:599-604. DOI: 10.1038/nphoton.2008.194
2. Brabec T, editor. Strong Field Laser Physics. New York: Springer; 2008. 585 p. DOI: 10.1007/978-0-387-34755-4
3. Marchese SV, Baer CRE, Engqvist AG, Hashimoto S, Maas DJHC, Golling M, Südmeyer T, Keller U. Femtosecond thin disk laser oscillator with pulse energy beyond the 10-microjoule level. Optics Express. 2008;16(9):6397-6407. DOI: 10.1364/OE.16.006397
4. Bauer D, Zawischa I, Sutter DH, Killi A, Dekorsy T. Mode-locked Yb:YAG thin-disk oscillator with 41 μJ pulse energy at 145 W average infrared power and high power frequency conversion. Optics Express. 2012;20(9):9698-9704. DOI: 10.1364/OE.20.009698
5. Saraceno CJ, Emaury F, Schriber C, Hoffmann M, Golling M, Südmeyer T, Keller U. Ultrafast thin-disk laser with 80 μJ pulse energy and 242 W of average power. Optics Letters. 2014;39(1):9-12. DOI: 10.1364/OL.39.000009
6. Saraceno CJ, Emaury F, Heckl OH, Baer CRE, Hoffmann M, Schriber C, Golling M, Südmeyer T, Keller U. 275 W average output power

from a femtosecond thin disk oscillator operated in a vacuum environment. Optics Express. 2012;20(21):23535-23541. DOI: 10.1364/OE.20.023535

7. Brons J, Pervak V, Fedulova E, Bauer D, Sutter D, Kalashnikov V, Apolonskiy A, Pronin O, Krausz F. Energy scaling of Kerr-lens mode-locked thin-disk oscillators. Optics Letters. 2014;39(22):6442-6445. DOI: 10.1364/OL.39.006442

8. Emaury F, Diebold A, Saraceno CJ, Keller U. Compact extreme ultraviolet source at megahertz pulse repetition rate with a low-noise ultrafast thin-disk laser oscillator. Optica. 2015;2(11):980-984. DOI: 10.1364/OPTICA.2.000980

9. Backus S, Durfee CG III, Murnane MM, Kapteyn HC. High power ultrafast lasers. Review of Scientific Instruments. 1998;69(3):1207-1223

10. Mourou GA, Tajima T, Bulanov SV. Optics in the relativistic regime. Reviews of Modern Physics. 2006;78(2):309-371. DOI: 10.1103/RevModPhys.78.309

11. Liu Y, Tschuch S, Rudenko A, Dürr M, Siegel M, Morgner U, Moshammer R, Ullrich J. Strong-field double ionization of Ar below the recollision threshold. Physical Review Letters. 2008;101:053001. DOI: 10.1103/PhysRevLett.101.053001

12. Seres E, Seres J, Spielmann C. Extreme ultraviolet light source based on intracavity high harmonic generation in a mode locked Ti:sapphire oscillator with 9.4 MHz repetition rate. Optics Express. 2012;20(6):6185-6190. DOI: 10.1364/OE.20.006185

13. Sciaini G, Miller RJD. Femtosecond electron diffraction: Heralding the era of atomically resolved dynamics. Reports on Progress in Physics. 2011;74:096101. DOI: 10.1088/0034-4885/74/9/096101

14. Gattass RR, Mazur E. Femtosecond laser micromachining in transparent materials. Nature Photonics. 2008;2:219-225. DOI: 10.1038/nphoton.2008.47

15. Lanin AA, Fedotov IV, Sidorov-Biryukov DA, Doronina-Amitonova LV, Ivashkina OI, Zots MA, Sun C-K, Ilday FÖ, Fedotov AB, Anokhin KV, Zheltikov AM. Air-guided photonic-crystal-fiber pulse-

compression delivery of multimegawatt femtosecond laser output for nonlinear-optical imaging and neurosurgery. Applied Physics Letters. 2012;100:101104. DOI: 10.1063/1.3681777

16. Gallmann L, Cirelli C, Keller U. Attosecond science: Recent highlights and future trends. Annual Review of Physical Chemistry. 2012;63:447-469. DOI: 10.1146/annurev-physchem-032511-143702
17. Akhmediev NN, Ankiewicz A, editors. Dissipative Soliton. Berlin: Springer; 2005. 456 p. DOI: 10.1007/b11728
18. Spatschek KH, Mertens FG, editors. Nonlinear Coherent Structures in Physics and Biology. New York: Springer; 1994. 476 p. DOI: 10.1007/978-1-4899-1343-2
19. Dawes JHP. The emergence of a coherent structure for coherent structures: Localized states in nonlinear systems. Philosophical Transactions of the Royal Society A. 2010;368:3519-3534. DOI: 10.1098/rsta.2010.0057
20. Kevrekidis PG, Frantzeskakis DJ, Carretero-Gonzalez R, editors. Emergent Nonlinear Phenomena in Bose-Einstein Condensates. Berlin: Springer; 2008. 385 p
21. Editorial. The power of analogues. Nature Photonics. 2014;8:1. DOI: 10.1038/nphoton.2013.359
22. Scott AC. The Nonlinear Universe: Chaos, Emergence, Life. Berlin: Springer; 2010. 364 p
23. Ph G, Akhmediev N. Dissipative solitons for mode-locked lasers. Nature Photonics. 2012;6:84-92. DOI: 10.1038/NPHOTON.2011.345
24. Kalashnikov VL. Chirped-pulse oscillators: Route to the energy scalable femtosecond pulses. In: Al-Khursan AH, editor. Solid-State Laser. Rijeka: InTechOpen; 2012. pp. 145-184. DOI: 10.5772/37415
25. Kalashnikov VL, Sergeyev SV. Dissipative solitons in fibre lasers. In: Paul MC, editor. Fibre Laser. Rijeka: InTechOpen; 2016. pp. 165-210. DOI: 10.5772/61713
26. Ashkin A, Boyd G, Dziedzic J. Resonant optical second harmonic generation and mixing. IEEE Journal of Quantum Electronics. 1966;2(6):109-124. DOI: 10.1109/JQE.1966.1074007

27. Yanovsky VP, Wise FW. Frequency doubling of 100-fs pulses with 50% efficiency by use of a resonant enhancement cavity. Optics Letters. 1994;19(23):1952-1954. DOI: 10.1364/OL.19.001952

28. Lilienfein N, Hofer C, Holzberger S, Matzer C, Zimmermann P, Trubetskov M, Pervak V, Pupeza I. Enhancement cavities for few-cycle pulses. Optics Letters. 2017;42(2):271-274. DOI: 10.1364/OL.42.000271

29. Siegman AE. Lasers. Mill Valley: University Science Books; 1986. 1283 p

30. Kalashnikov VL. Mathematical Ultrashort-Pulse Laser Physics [Internet]. March 26, 2002. Available from: https://www.maplesoft.com/applications/view.aspx?SID=4247

31. Abraham NB, Lugiato LA, Narducci LM. Overview of instabilities in laser systems. Journal of the Optical Society of America B: Optical Physics. 1985;2(1):7-14. DOI: 10.1364/JOSAB.2.000007

32. Goodberlet J, Wang J, Fujimoto JG, Schulz PA. Starting dynamics of additive-pulse mode locking in the Ti:Al_2O_3 laser. Optics Letters. 1990;15(22):1300-1302. DOI: 10.1364/OL.15.001300

33. Krausz F, Fermann ME, Brabec T, Curley PF, Hofer M, Ober MH, Spielmann C, Wintner E, Schmidt AJ. Femtosecond solid-state lasers. IEEE Journal of Quantum Electronics. 1992;28(10):2097-2122. DOI: 10.1109/3.159520

34. Krausz F, Brabec T. Passive mode locking in standing-wave laser resonators. Optics Letters. 1993;18(11):888-890. DOI: 10.1364/OL.18.000888

35. Ippen EP, Liu LY, Haus HA. Self-starting condition for additive-pulse mode-locked lasers. Optics Letters. 1990;15(3):183-185. DOI: 10.1364/OL.15.000183

36. Komarov AK, Komarov KP, Mitschke FM. Phase-modulation bistability and threshold self-start of laser passive mode locking. Physical Review A. 2002;65:053803. DOI: 10.1103/PhysRevA.65.053803

37. Chen C-J, Wai PKA, Menyuk CR. Self-starting of passively mode-locked lasers with fast saturable absorbers. Optics Letters. 1995;20(4):350-352. DOI: 10.1364/OL.20.000350

38. Soto-Crespo JM, Akhmediev N, Town G. Continuous-wave versus pulse regime in a passively mode-locked laser with a fast saturable absorber. Journal of the Optical Society of America B. 2002;19(2):234-242. DOI: 10.1364/JOSAB.19.000234

39. Risken H, Nummedal K. Self-pulsing in lasers. Journal of Applied Physics. 1968;39(10):4662-4672. DOI: 10.1063/1.1655817

40. Gordon A, Fischer B. Phase transition theory of pulse formation in passively mode-locked lasers with dispersion and Kerr nonlinearity. Optics Communications. 2003;223:151-156. DOI: 10.1016/S0030-4018(03)01622-5

41. Vodonos B, Bekker A, Smulakovsky V, Gordon A, Gat O, Berger NK, Fischer B. Experimental study of the stochastic nature of the pulsation self-starting process in passive mode locking. Optics Letters. 2005;30(20):2787-2789. DOI: 10.1364/OL.30.002787

42. Gordon A, Gat O, Fischer B, Kärtner FX. Self-starting of passive mode locking. Optics Express. 2006;14(23):11142-11154. DOI: 10.1364/OE.14.011142

43. Gammaitoni L, Hänggi P, Jung P, Marchesoni F. Stochastic resonance. Reviews of Modern Physics. 1998;70(1):223-287

44. Kivshar YS, Agrawal GP. Optical Solitons: From Fibers to Photonic Crystals. Amsterdam: Academic Press; 2003. 540 p

45. Aranson IS, Kramer L. The world of the complex Ginzburg-Landau equation. Reviews of Modern Physics. 2002;79(1):99-143. DOI: 10.1103/RevModPhys.74.99

46. García-Morales V, Krischer K. The complex Ginzburg-Landau equation: An introduction. Contemporary Physics. 2012;53(2):79-95. DOI: 10.1080/00107514.2011.642554

47. van Saarloos W, Hohenberg PC. Fronts, pulses, sources and sinks in generalized complex Ginzburg-Landau equations. Physica D. 1992;56:303-367

48. Akhmediev NN, Ankiewicz A, editors. Dissipative Solitons: from Optics to Biology and Medicine. Berlin: Springer; 2008

49. Anderson D, Lisak M, Berntson A. A variational approach to nonlinear evolution equations in optics. Pramana-Journal of Physics. 2001;57(5–6):917-936. DOI: 10.1007/s12043-001-0006-z

50. Malomed BA. Variational methods in nonlinear fiber optics and related fields. In: Wolf E, editor. Progress in Optics. Amsterdam: Elsevier; 2002. pp. 71-193

51. Ankiewicz A, Akhmediev N, Devine N. Dissipative solitons with a Lagrangian approach. Optical Fiber Technology. 2007;13(2):91-97. DOI: 10.1016/j.yofte.2006.12.001

52. Kalashnikov VL. Chirp-free Solitons in Dissipative Systems: Variational Approximation and Issue of the Soliton Energy Scalability [Internet]. May 30, 2014. Available from: http://info.tuwien.ac.at/kalashnikov/variational_soliton.pdf

53. Kantorovich LV, Krylov VI. Approximate Methods of Higher Analysis. Groningen: P. Noordhoff Ltd; 1958. 681 p

54. Cerda SC, Cavalcanti SB, Hickmann JM. A variational approach to nonlinear dissipative pulse propagation. European Physical Journal D: Atomic, Molecular, Optical and Plasma Physics. 1998;1:313-316

55. Kalashnikov VL, Sorokin E, Sorokina IT. Multipulse operation and limits of the Kerr-lens mode locking stability. IEEE Journal of Quantum Electronics. 2003;39(2):323-336. DOI: 10.1109/JQE.2002.807204

56. Vakhitov NG, Kolokolov AA. Stationary solutions of the wave equation in a medium with nonlinearity saturation. Radiophysics and Quantum Electronics. 1973;16(7):783-789

57. Bergé L. Wave collapse in physics: Principles and applications to light and plasma waves. Physics Reports. 1998;303(5–6):259-370. DOI: 10.1016/S0370-1573(97)00092-6

58. Baer CRE, Heckl OH, Saraceno CJ, Schriber C, Kränkel C, Südmeyer T, Keller U. Frontiers in passively mode-locked high-power thin disk laser oscillators. Optics Express. 2012;20(7):7054-7065. DOI: 10.1364/OE.20.007054

59. Saraceno CJ, Heckl OH, Baer CRE, Schriber C, Golling M, Beil K, Kränkel C, Südmeyer T, Huber G, Keller U. Sub-100 femtosecond pulses from a SESAM modelocked thin disk laser. Applied Physics B. 2012;106(3):559-562. DOI: 10.1007/s00340-012-4900-5

60. Paré C, Bélanger P-A. Optical solitary waves in the presence of a Lorentzian gain line: Limitations of the Ginzburg–Landau model. Optics Communications. 1998;145(1–6):385-392. DOI: 10.1016/S0030-4018(97)00433-1

61. Kalashnikov VL, Apolonski A. Energy scalability of mode-locked oscillators: A completely analytical approach to analysis. Optics Express. 2010;18(25):25757-25770. DOI: 10.1364/OE.18.025757

62. Zhang J, Brons J, Seidel M, Pervak V, Kalashnikov V, Wei Zh, Apolonski A, Krausz F, Pronin O. 49-fs Yb: YAG thin-disk oscillator with distributed Kerr-lens mode-locking. In: European Conference on Lasers and Electro-Optics—European Quantum Electronics Conference; 21–25 June 2015; Munich, Germany. Optical Society of America; 2015. p. PD_A_1

63. Podivilov E, Kalashnikov VL. Heavily-chirped solitary pulses in the normal dispersion region: New solutions of the cubic-quintic complex Ginzburg-Landau equation. JETP Letters. 2005;82(8):467-471

64. Kalashnikov VL. Chirped-pulse oscillators: A unified standpoint. Physical Review A. 2009;79:043829. DOI: 10.1103/PhysRevA.79.043829

65. Chang W, Ankiewicz A, Soto-Crespo JM, Akhmediev N. Dissipative soliton resonances. Physical Review A. 2008;78:023830. DOI: 10.1103/PhysRevA.78.023830

66. Kalashnikov VL. Chirped Solitary-Pulse Solutions of the Complex Generalized Nonlinear Ginzburg-Landau Equation (II) [Internet]. February 2, 2009. Available from: http://info.tuwien.ac.at/kalashnikov/genNCGLE.html

67. Kalashnikov VL, Sorokin E, Sorokina IT. Chirped dissipative soliton absorption spectroscopy. Optics Express. 2011;19(18):17480-17492

68. Zhu L, Verhoef AJ, Jespersen KG, Kalashnikov VL, Grüner-Nielsen L, Lorenc D, Baltuška A, Fernández A. Generation of high fidelity 62-fs, 7-nJ pulses at 1035 nm from a net normal-dispersion Yb-fiber laser with anomalous dispersion higher-order-mode fiber. Optics Express. 2013;21(14):16255-16262. DOI: 10.1364/OE.21.016255

69. Pronin O, Brons J, Grasse C, Pervak V, Boehm G, Amann M-C, Apolonski A, Kalashnikov VL, Krausz F. High-power Kerr-lens mode-locked Yb:YAG thin-disk oscillator in the positive dispersion regime. Optics Letters. 2012;37(17):3543-3545. DOI: 10.1364/OL.37.003543

70. Li H, Ouzounov DG, Wise FW. Starting dynamics of dissipative-soliton fiber laser. Optics Letters. 2010;35(14):2403-2405. DOI: 10.1364/OL.35.002403

71. Kalashnikov VL, Kalosha VP, Poloiko IG, Mikhailov VP. New principle of formation of ultrashort pulses in solid-state lasers with self-phase-modulation and gain saturation. Quantum Electronics. 1996;26(3):236-239

72. Connaughton C, Josserand C, Picozzi A, Pomeau Y, Rica S. Condensation of classical nonlinear waves. Physical Review Letters. 2005;95:263901. DOI: 10.1103/PhysRevLett.95.263901

73. Michel C, Kibler B, Picozzi A. Discrete spectral incoherent solitons in nonlinear media with noninstantaneous response. Physical Review A. 2011;83:023806. DOI: 10.1103/PhysRevA.83.023806

74. Kalashnikov VL. Optics and chaos: Chaotic, rogue, and noisy optical dissipative solitons. In: Skiadas CH, Skiadas C, editors. Handbook of Applications of Chaos Theory. London: CRC Press; 2016. pp. 587-626

75. Kalashnikov VL, Podivilov E, Chernykh A, Naumov S, Fernandez A, Graf R, Apolonski A. Approaching the microjoule frontier with femtosecond laser oscillators: Theory and comparison with experiment. New Journal of Physics. 2005;7:217. DOI: 10.1088/1367-2630/7/1/217

76. Kalashnikov VL, Podivilov E, Chernykh A, Apolonski A. Chirped-pulse oscillators: Theory and experiment. Applied Physics B. 2006;83:503-510. DOI: 10.1007/s00340-006-2214-1

77. Kalashnikov VL. Chirped-Pulse Oscillators: An Impact of the Dynamic Gain Saturation [Internet]. July 7, 2008. Available from: https://arxiv.org/abs/0807.1050

78. Kharenko DS, Podivilov EV, Apolonski AA, Babin SA. 20 nJ 200 fs all-fiber highly chirped dissipative soliton oscillator. Optics Letters. 2012;37(19):4104-4106. DOI: 10.1364/OL.37.004104

79. Babin SA, Podivilov EV, Kharenko DS, Bednyakova AE, Fedoruk MP, Kalashnikov VL, Apolonski A. Multicolour nonlinearly bound chirped dissipative solitons. Nature Communications. 2014;5:4653. DOI: 10.1038/ncomms5653

80. Kalashnikov VL, Evgeni S. Dissipative Raman solitons. Optics Express. 2014;22(24):30118-30126. DOI: 10.1364/OE.22.030118

81. Babin SA, Podivilov EV, Kharenko DS, Bednyakova AE, Fedoruk MP, Shtyrina OV, Kalashnikov VL, Apolonski AA. SRS-driven evolution of dissipative solitons in fiber lasers. In: Ph G, editor. Nonlinear Optical Cavity Dynamics: From Microresonators to Fiber Lasers. Mannheim: Wiley; 2016. pp. 277-316. DOI: 10.1002/9783527686476.ch12

82. Kalashnikov VL, Sorokin E. Turbulence of optical dissipative solitons. In: Skiadas ChH, editor. Book of Abstracts of 10th Chaotic Modeling and Simulation International Conference; 30 May–2 June, 2017; Barcelona, Spain. ISAST: International Society for the Advancement of Science and Technology; 2017. p. 64

83. Bernhardt B, Ozawa A, Jacquet P, Jacquey M, Kobayashi Y, Udem T, Holzwarth R, Guelachvili G, Hänsch TW, Picqué N. Cavity-enhanced dual-comb spectroscopy. Nature Photonics. 2010;4:55-57. DOI: 10.1038/nphoton.2009.217

84. Pupeza I, Holzberger S, Eidam T, Carstens H, Esser D, Weitenberg J, Rußbüldt P, Rauschenberger J, Limpert J, Udem T, Tünnermann A, Hänsch TW, Apolonski A, Krausz F, Fill E. Compact high-repetition-rate source of coherent 100 eV radiation. Nature Photonics. 2013;7:608-612. DOI: 10.1038/nphoton.2013.156

85. Carstens H, Högner M, Saule T, Holzberger S, Lilienfein N, Guggenmos A, Jocher C, Eidam T, Esser D, Tosa V, Pervak V, Limpert

J, Tünnermann A, Kleineberg U, Krausz F, Pupeza I. High-harmonic generation at 250 MHz with photon energies exceeding 100 eV. Optica. 2016;3(4):366-369. DOI: 10.1364/OPTICA.3.000366

86. Ippen EP, Haus HA, Liu LY. Additive pulse mode locking. Journal of the Optical Society of America B. 1989;6(9):1736-1745. DOI: 10.1364/JOSAB.6.001736

87. Kalashnikov VL. Femtosecond pulse enhancement in an external resonator: Impact of dispersive and nonlinear effects. Applied Physics B: Lasers and Optics. 2008;92:19-23. DOI: 10.1007/s00340-008-3064-9

88. Kalashnikov VL, Gohle Ch, Udem Th. Maximization of ultrashort pulse power stored in a passive resonator synchronously pumped by a femtosecond oscillator. In: Denman C, Sorokina I, editors. Advanced Solid-State Photonics (TOPS); February 6–9, 2005; Vienna, Austria. Optical Society of America; 2005. p. 652. DOI: 10.1364/ASSP.2005.652

FOOTNOTES

[1]This phenomenon is closely related to the concept of *stochastic resonance* which describes processes of resonant coherence enhancement in a noisy periodically driven system [43].

[2]Different versions of this equation describe an extremely broad area of phenomena ranging from laser dynamics [17, 46, 47], oscillatory chemical reactions [22] to Bose-Einstein condensations and biological systems [48].

[3]In the case of perfectly saturable SAM, the confinement is imposed by spectral dissipation (i.e., the DS width but not power is confined; see Eq. (11)).

[4]The negative sign of β_2 corresponds to a normal GDD in these notations.

[5]The adiabatic theory does not predict a spectral condensation near the carrier frequency $\omega \approx 0$ for this SAM law, but such a concentration is possible at *spectrum edges*. This phenomenon is clearly visible in the numerous experiments and can be explained by the DS perturbation theory [67].

[6]That is a measure of the DS *fidelity*.

[7]$\Gamma \ll 1$ and $m \leq 4$ are assumed. The field amplitude and the pulse width are normalized to those of incident.

[8]Here $\beta_2 < 0$ corresponds to an *anomalous* dispersion.

[9]$r < \Gamma$ is the stability condition.

CHAPTER-10

PLASMA-MEDIATED SELF-FOCUSING OF A HIGH-POWER LASER BEAM

1. INTRODUCTION: AN OVERVIEW

The self-focusing phenomena is regarded as a significant self-action effect in the interaction between lasers and plasmas [1, 2]. During this process, the laser beams undergo a nonlinear reaction of plasma, which allows them to adjust their front medium in order to enhance their suitability for propagation. In this situation, the refractive index of plasma could be expressed as $n = n_0 + n_2 I r$ in which n 0 presents a linear term, n_2 is considered as an optical constant characterizing the strength of optical nonlinearity, and I r determines the beam intensity distribution along the radial coordinate r . Put simply, when a powerful laser beam with varying intensity over its width moves through plasma, the laser beam often experiences self-focusing and self-defocusing. If the electric field reaches a sufficient strength, the laser beam will generate a dielectric waveguide in the upcoming route. This conventional waveguide leads to the reduction or complete elimination of beam divergence. Optically, the refractive index of the medium in these cases behaves like a convex lens. As a result, the middle portion of the laser beam moves at a slower pace compared to the outside regions. Thus, when the beam travels through the nonlinear medium, its wave front progressively distorts, as shown in Figure 1.

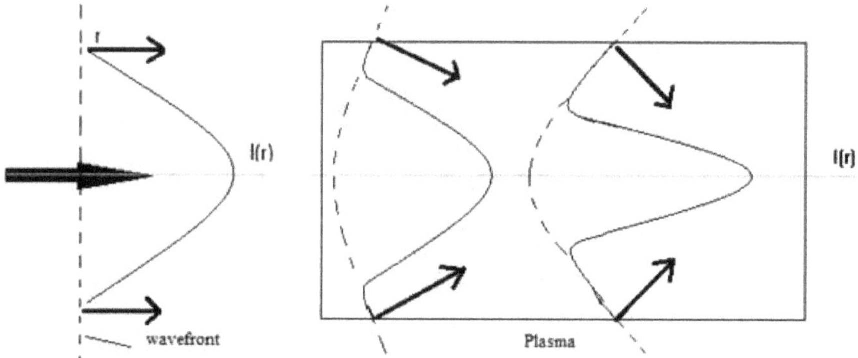

Figure 1. A schematic showing distortion of the wave-front and self-focusing of a laser beam in plasma.

Typically, the self-focusing phenomena arises due to various physical causes. The self-focusing of a laser beam is a result of the nonlinearity of the medium, which arises from its interaction with the laser field. Therefore, the process of a laser beam concentrating itself through plasma can be categorized into three distinct groups according to the nonlinear mechanics involved. These categories are listed below:

1. Thermal self-focusing (TS)

This effect is due to collisional heating of plasma exposed to electromagnetic radiation. In fact, the rise in temperature induces the hydrodynamic expansion, which leads to an increase in refractive index and further heating [3].

2. Ponderomotive self-focusing (PS)

A nonlinear radial ponderomotive force of the focused laser beam pushes electrons out of the propagation axis. It expels the plasma from the beam centre, high-intensity region, and increases the plasma dielectric function, leading to self-focusing of the laser in plasmas.

3. Relativistic self-focusing (RS)

The increase of electrons' mass traveling by velocity approaching the speed of light modifies the plasma refractive index. This phenomenon has been observed in several experiments and has been proved to be an efficient way to guide a laser pulse over distances much longer than the Rayleigh length.

R.W. Boyd et al. [4] reviewed the self-focusing methods, which are recommended by the authors for more details on the topic.

According to the self-focusing of laser beam through plasma, the irradiance of a focused pulse laser with a range of 10^{15} - 10^{16}W/cm^2, for instance, can reach approximately 10^{18}W/cm^2. Under this high intensity, the relativistic motion of the electrons may be expected. This mechanism is of great importance because it is able to produce ultra-high laser irradiances (10^{19} - 10^{20}W/cm^2) over distances much greater than the Rayleigh length [5]. Sun et al. [6], Barnes et al. [7], Kurki-Suonio et al. [8], and Borisov et al. [9] investigated both relativistic mass variation and charge displacement due to the ponderomotive force by solving the propagation within the envelope approximation and fluid equations. With advancement in laser technology using the chirped-pulse amplification technique [10, 11], the propagation of laser beam in inhomogeneous plasmas has been also studied [12, 13]. Furthermore, guiding nonlinear relativistic laser pulses in preformed plasma with density transition has recently become a subject of great interests [14, 15, 16, 17].

A fundamental example of a nonlinear partial differential equation in physics is the self-focusing wave equation. It serves as a prototype for an important class of equations, including the Landau Ginsberg equation for the macroscopic wave function of type II superconductors and Schrodinger's equation for a particle with self-interactions [18]. There are multiple approximate analytical methods available to analyze the impact of self-focusing in this scenario. These include the paraxial ray approximation, moment theory approach, variation technique, and source-dependent expansion method. Nevertheless, all of these ideas have constraints when it comes to fully elucidating the experimental and computer simulation findings.

The discussion commences by examining the equations commonly employed to depict self-focusing in the Wentzel-Kramers-Brillouin (WKB) and paraxial ray approximation (Section 2). This is followed by an overview of the ramp-density profile and its influence on the self-focusing of high-intensity laser beams. The following sections discuss the examination of relativistic self-focusing in classical and quantum plasma using Gaussian and Cosh-Gaussian laser beams.

2. SELF-FOCUSING EQUATION FOR HIGH-INTENSE LASER-PLASMA INTERACTION

2.1. High-order paraxial theory

It is reasonable to assume that the paraxial wave equation presents an accurate description for laser beams propagating near the axis throughout the propagation. Akhmanove et al. [19] illustrated that in a limit when the eikonal term is expanded only up to the second power in r, the shape of the radial intensity profile remains unchanged. However, in the experimental situation with high-intense laser beams, one needs to go beyond the paraxial approximation for which the predictions of such an approximation are often not sufficiently accurate [20]. Thus, it would be interesting, on high-intense laser-plasma interaction, to investigate propagation of laser beams using the extended paraxial approximation. In this case, Liu [21] and Tripathi [22] reported a useful theoretical framework that accounts for the combined several effects of interaction of an intense short pulse laser with plasma, the laser frequency blue shift, self-defocusing, ring formation and self-phase modulation. The expansion of the eikonal term to the fourth power in r makes significant difference in studying laser beam propagating through plasma and even other nonlinear materials.

For more details, the interaction of an intense laser beam with particular plasma is considered. From Maxwell's equations, it is noted that the propagation of such a beam can be investigated by solving the scalar wave equation in the cylindrical coordinate system and along the axis z:

$$\frac{\partial^2}{\partial z^2}E(r,z,t) + \nabla_\perp^2 E(r,z,t) - \frac{\varepsilon}{c^2}\frac{\partial^2}{\partial t^2}E(r,z,t) = 0 \tag{1}$$

where E shows amplitude of the electric field, c is the speed of light in vacuum and ε is the dielectric constant of plasma. In this stage, we consider the solution of the Eq. (1):

$$E(r,z,t) = A(r,z,t)\exp\left[i\left(\omega t - \int_0^z k(z)dz\right)\right] \tag{2}$$

E(r, z, t) can be substituted with Eq. (2) in Eq. (1). This substitution leads to neglecting $\partial^2 A/\partial z^2$ and $\partial^2 A/\partial t^2$ on the assumption that A is a slowly varying function of z and t is compared with ω:

$$-2ik\frac{\partial A}{\partial z} - iA\frac{\partial k}{\partial z} - k^2 A + \frac{\partial^2 A}{\partial r^2} + \frac{1}{r}\frac{\partial A}{\partial r} + \frac{\omega^2}{c^2}\varepsilon A = 0. \tag{3}$$

The complex amplitude of the electric vector $A(r, z, t)$ is expressed as,

$$A(r,z,t) = A_0(r,z,t)\exp\left[-ik(z)S(r,z)\right] \tag{4}$$

From Eq. (4), it is noticed that the envelope $A(r, z, t)$ has been separated into real amplitude and complex phase terms in which the eikonal function $S(r, z)$ is.

$$S(r,z) = S_0(z) + (r^2/r_0^2)S_2(z) + (r^4/r_0^4)S_4(z) \tag{5}$$

where $S_0(z)$ is the axial phase shift, $S_2(z) = (df(z)/dz)/2f(z)$ is indicative of the spherical curvature of the wave front, and $S_4(z)$ corresponds to its departure from the spherical nature. Moreover, the beam irradiance $A_0(r, z, t)$ of Cosh-Gaussian (ChG) laser beam can be written as:

$$\begin{aligned} A_0^2(r,z,t) = EE^* &= \frac{E_0^2}{4f^2(z)}\exp\left(\frac{b^2}{2}\right)\left(1 + \frac{r^2}{r_0^2 f^2(z)}a_2(z) + \frac{r^4}{r_0^4 f^4(z)}a_4(z)\right) \\ &\times \left(\exp\left[-\left(\frac{r}{r_0 f(z)} + \frac{b}{2}\right)^2\right] + \exp\left[-\left(\frac{r}{r_0 f(z)} - \frac{b}{2}\right)^2\right]\right)^2 g(t) \end{aligned} \tag{6}$$

In the near-axis approximation (i.e. a 2 , a 4 → 0), Eq. (6) converts to a general solution of ChG laser beam. In Eq. (6), the initial laser intensity at the central position r = z = 0 is presented by E_0, and the beam-width parameter characterized by *f*(z) depends on z. Temporal shape of the pulse can be considered as a step function, g(t) = 1 at t〉0 and g(t) = 0 otherwise. In addition, b is called the decentred parameter as well as r 0 introduced the initial spot size of the laser beam. Two functions $a_2(z)$ and $a_4(z)$, the coefficients of r^2 and r^4, respectively, are considered as an indicative of the departure of the beam from the Gaussian nature.

Furthermore, in the higher-order paraxial theory, the dielectric constant is expanded to the next higher power in r^2 to obtain $\varepsilon(r, z) = \varepsilon_0(z) - (r^2/r_0^2)\ \varepsilon_2(z) - (r^4/r_0^4)\ \varepsilon_4 z$. The parameters $\varepsilon_0(z)$, $\varepsilon_2(z)$, and $\varepsilon_4(z)$ corresponding to the nonlinearity play an important role in investigating the self-focusing of laser beam through a nonlinear medium. By substituting both Eqs. (5) and (6) into Eq. (3) and equating the coefficients of r^0, r^2, and r^4 on both sides of the resulting equation, the differential equations governing the parameters a^2, a^4, $f(z)$, and $S_4(z)$ can be given by:

$$\frac{da_2}{d\xi} = -16S_4'f^2, \quad a_4 = (3a_2^2 - 4a_2)/4 \tag{7}$$

$$\frac{d^2f}{d\xi^2} = \frac{\{(3a_2^2 - 8a_2 + 4) - b^2(4 + b^2/3 - 2a_2)\}}{\varepsilon_0 f^3} - \frac{\varepsilon_2}{\varepsilon_0}\rho^2 f - \frac{1}{2\varepsilon_0}\frac{d\varepsilon_0}{d\xi}\frac{df}{d\xi} \tag{8}$$

$$\frac{dS_4'}{d\xi} = \frac{40a_2^2 - 52a_2^3 + 32a_2}{8\varepsilon_0 f^6} + \frac{(a_2^2 + (4 - 8a_2)/3 + 2b^6/15)b^2}{\varepsilon_0 f^6} - \frac{\varepsilon_4}{2\varepsilon_0}\rho^2 - \frac{S_4'}{2\varepsilon_0}\frac{d\varepsilon_0}{d\xi} - \frac{4S_4'}{f}\frac{df}{d\xi} \tag{9}$$

where $S_4' = S_4\omega/c$. Additionally, $\xi = (c/\omega r_0^2)z$ and $\rho = r_0\omega/c$ are considered as a dimensionless propagation distance and an original beam-width parameter, respectively. The first term on the right-hand side of Eq. (8) represents the diffraction effect, while the second term plays a vital role in self-focusing of laser beam. Both of them are nonlinear and responsible for the defocusing and focusing of the ChG laser beam through plasma. The behaviour of laser beam propagating through plasma can be investigated by solving these equations, Eqs. (7)–(9), with the initial conditions $f = 1$, $df/d\xi = 0$, $S_4' = 0$ and $a_2 = 0$ at $\xi = 0$. In non-stationary situations, the time derivative is removed from the wave equation using the independent variable transformation ($Z = z$, $T = t - z/vg$) where v g is group velocity. We have presented non-stationary self-focusing of high-intense Gaussian laser beams for different portions of a pulse in classical and quantum plasmas in the weakly relativistic as well as ponderomotive regimes [23, 24]. We have recorded a very significant focusing near the peak of the pulse and the rear portion of the pulse.

2.2. Importance of nonlinearity in self-focusing equation

As mentioned in the previous section, the refractive index of plasma, the second term in the right-hand side of the self-focusing equation, Eq. (8), depends on nonlinearity mechanisms. Therefore, this term plays an important role in investigation propagation laser beam through a nonlinear medium with a wide variety of nonlinearities. For example, in collisional nonlinearity condition:

$$\varepsilon(r, z) = 1 - \frac{\omega_p^2}{\omega^2} + \frac{\omega_p^2}{\omega^2}\frac{\alpha EE^*/2T_e}{1 + \alpha EE^*/2T_e}, \tag{10}$$

where $\alpha = e^2/3\delta m\omega^2 kB$

or in the ponderomotive regime and in low-intensity laser, it should be expressed like,

$$\varepsilon(r,z) = 1 - \frac{\omega_p^2}{\omega^2} + \frac{\omega_p^2}{\omega^2}\left[1 - \exp\left(-\frac{\beta EE^*}{T_e}\right)\right], \tag{11}$$

where $\beta = \frac{e^2}{8m\omega^2 k_B}$

In addition, in the relativistic regime and high-intense laser-plasma interaction:

$$\varepsilon(r,z) = 1 - \frac{1}{\gamma}\frac{\omega_p^2}{\omega^2}, \gamma \simeq \left(1 + a_\circ^2/2\right)^{1/2} \text{ Circular Polarization, } a_\circ = \frac{eE_\circ}{\omega mc} \tag{12}$$

where γ is the Lorentz factor which arises from the quiver motion of the electron in the laser field. The expressions earlier are just three forms of nonlinearities in plasma. In this case, we have investigated relativistic self-focusing of high-intense laser beam in cold and warm quantum plasma [25, 26, 27, 28]. From a quantum-mechanical viewpoint, the de Broglie wavelength of the charge particle is comparable to the inter-particle distance. In this situation, the dielectric constant in relativistic cold quantum plasma (CQP) is expressed by:

$$\varepsilon(r,z) = 1 - \frac{1}{\gamma}\frac{\omega_p^2}{\omega^2}\left(1 - \hbar^2 k^4/4\gamma m_0{}^2\omega^2\right)^{-1} \tag{13}$$

and in the relativistic warm quantum plasma:

$$\varepsilon(r,z) = 1 - \frac{1}{\gamma}\frac{\omega_p^2}{\omega^2}\left(1 - 2k^2 k_B T_{Fe}/m_0{}^2\omega^2 - \hbar^2 k^4/4\gamma m_0{}^2\omega^2\right)^{-1} \tag{14}$$

2.3. Ramp density profile

The plasma density profile is another crucial component in solving the self-focusing equation. Mathematically and practically, it can be viewed as a function of propagation distance that can either be uniform or non-uniform. The study examined the behavior of a Gaussian high-intensity laser beam as it propagated through under-dense plasma with a density ramp that increased in an upward direction, in the context of inhomogeneous plasmas [12]. This chapter also demonstrates the impact of electron density profile on the oscillations of spot size in a laser beam. It results in additional variations in the measurement of the laser beam's spot size. Thus, it has been verified that an enhanced electron density gradient profile is a crucial determinant for achieving effective stationary and non-stationary self-focusing in laser-plasma interaction. An equation that represents the distribution of charges in a non-uniform manner, used to model plasma that is not evenly distributed, can be expressed as:

$$n_e(\xi) = n_0 F(\xi/d) \tag{15}$$

where $\xi = z/R_d$ is a dimensionless propagation distance, $R_d = \omega r_0^2/c$ is the Rayleigh length, r_0 is the focused laser beam radius, n_0 is the density of the plasma at $\xi = 0$, and $F(\xi) = 1 + (n_1/n_0)\tan \xi/d$ is the linear density profile function. The gradient of the ramp density profile can be ascertained by modifying the values of the parameters d and n1. The density model described here is a specific type of plasma density transition known as the ramp density profile. This model is used to study several laser-plasma phenomena, including self-focusing of laser beams, electron acceleration, and high harmonic production. This mathematical model of density change has also received support from numerous numerical and experimental studies [29, 30, 31, 32, 33, 34, 35]. For instance, Chunyang et al. [36] detected the presence of high harmonics in the reflection spectra resulting from the interaction of short-intense laser pulses with over-dense plasmas using particle-in-cell (PIC) simulations.

3. RELATIVISTIC SELF-FOCUSING OF CHG LASER BEAM IN QUANTUM PLASMA

Quantum effects on the propagation of Gaussian laser beams in dense plasmas are studied using a theoretical method [37, 38, 39, 40]. In their study, Habibi et al. demonstrated the significant impact of the Fermi temperature on enhancing the relativistic self-focusing of a short-wavelength laser beam (X-ray) through warm quantum plasmas [26]. The relativistic effect would be observed in dense degenerate plasmas due to an increase in the number density of fermions. Nevertheless, a number of recent advancements in technology have enabled the production of plasmas that possess densities comparable to that of solid state materials. Recently, there has been significant interest in the development and propagation of a decentred Gaussian beam. This is because these beams have a higher power efficiency and a flat-top beam shape compared to Gaussian laser beams. They also have appealing uses in complicated optical systems. In general, the analysis of the focusing of the ChG beam can be treated similarly to that of a Gaussian beam in plasmas, without taking into account quantum effects. This section focuses on examining the nonlinear propagation of a ChG laser beam in quantum plasma, specifically using higher-order paraxial theory.

The ChG laser beam exhibits a significant impact, surpassing the self-focusing effect of a Gaussian laser beam, when interacting with cold quantum plasma (CQP).

In this chapter, the plasma dielectric function, Eq. (13), which is in the relativistic regime, is considered for unmagnetized and collisionless CQP. Then, it is expanded to the next higher power in r^4 to obtain $\varepsilon(r, z) = \varepsilon_0(z) - (r^2/r_0^2)\,\varepsilon_2(z) - (r^4/r_0^4)\,\varepsilon_4(z)$. The parameters $\varepsilon_0(z)$, $\varepsilon_2(z)$, and $\varepsilon_4(z)$ corresponding to the relativistic nonlinearity are:

$$\varepsilon_0(z) = 1 - \frac{\omega_p^2}{\omega^2}\left(1+\frac{\Gamma}{f^2(z)}\right)^{-\frac{1}{2}}\left[1-\hbar^2k^4\left(1+\frac{\Gamma}{f^2(z)}\right)^{-1/2}/4{m_0}^2\omega^2\right]^{-1} \tag{16}$$

$$\varepsilon_2(z) = -\frac{\Gamma(-2+b^2+a_2(z))}{2f^4(z)}\frac{\omega_p^2}{\omega^2}\left(1+\frac{\Gamma}{f^2(z)}\right)^{-\frac{3}{2}}\left[1-\hbar^2k^4\left(1+\frac{\Gamma}{f^2(z)}\right)^{-1/2}/4{m_0}^2\omega^2\right]^{-2} \tag{17}$$

$$\varepsilon_4(z) = -\frac{\Gamma(\beta_1 f^2(z)+\Gamma\beta_2/4)}{6f^8(z)}\frac{\omega_p^2}{\omega^2}\left(1+\frac{\Gamma}{f^2(z)}\right)^{-\frac{5}{2}} \times\left[-2\beta_1 f^2(z)+3\Gamma(a_2^2(z)-2a_4(z))+1+4b^4+\hbar^2k^4\left(1+\frac{\Gamma}{f^2(z)}\right)^{-1/2}/4{m_0}^2\omega^2\right] \tag{18}$$

where $\Gamma = \alpha E_0^2$, $\beta_1 = 6 - 6b^2 + b^4 + 3a_4 - 6a_2 + 3a_2b^2$, and $\beta_2 = -3a_2^2 + (6b^2 - 12)a_2 - 12b^2 + b^4 + 12a_4 + 12$. By substituting these expressions in Eq. (8), the relativistic self-focusing of ChG laser beam through cold quantum plasma could be investigated. Eqs. (7)–(9) are numerically solved simultaneously using the fourth-order Runge-Kutta method with the initial conditions $f\,0 = 1$, $f'(0) = 0$, $S_4' = 0$, $a_2 = 0$ at $\xi = 0$ as well as following the set of parameters: $\Gamma = 0.08, 0.12, 0.14$, $r_o = 20$ μm, and $\omega_p = 0.59\,\omega$, $\omega = 1.77 \times 10^{16}\ s^{-1}$ which correspond to the gold metallic plasma at room temperature. It is noted that the case of a Gaussian beam b = 0.0 is similar to a dark ring, maximum irradiance on the axis.

Therefore, no parts of the beam penetrate beyond the determined depth of penetration. While in the case of a ChG beam, where b ≠ 0.0 like a bright ring, the irradiance is maximum on a ring and hence the portion of the beam around the bright ring is able to penetrate farther than that of on the axis. As a result, the decentred parameter plays an effective role in improving relativistic self-focusing of high-power Cosh-Gaussian laser beams in quantum plasma.

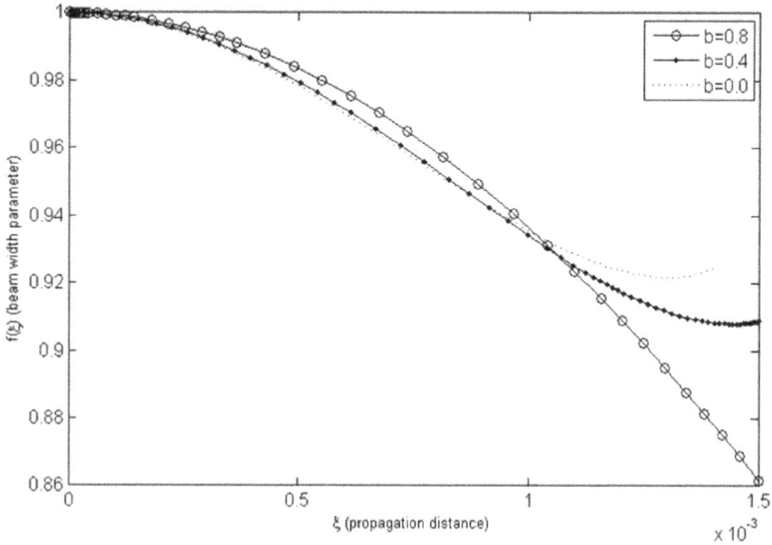

Figure 2. Variation of beam-width parameter f with the normalized propagation distance ξ for different values of decentred parameters b = 0.0, 0.4, 0.8 and Γ = 0.12, r_0 = 20 μm, ω_p = 0.59 ω.

Therefore, a comparative analysis can be done among various decentred parameters so as to determine its role in relativistic self-focusing of laser beam. At the first step, the variation of f on ξ in both Gaussian and Cosh-Gaussian of intensity distribution through the Q-plasma for three decentred parameters is shown in Figure 2. As seen from this figure, the focusing term becomes dominant with an increasing value of the decentred parameter b. The self-focusing effect is stronger for a higher decentred parameter at b = 0.8 than Gaussian laser beam b = 0. Consequently, increase of decentred parameters in the ChG laser beams will result in better reduction in the focusing length and more enhancements in localization of non-Gaussian laser beam as it is moving through plasma. For more investigation and better comparison, the variation of beam-width parameter with propagation distance for various powers of pump laser beam with b = 0.8 and through plasma density $\omega_p = 0.75\omega$ is shown in Figure 3. It is obvious that the laser beam focused substantially when the initial power of the laser beam grew from 0.08 to 0.14. As a result, the intensity of laser also plays an important role in enhancing the focusing of ChG laser beam through the cold dense plasmas along with the decentred parameter.

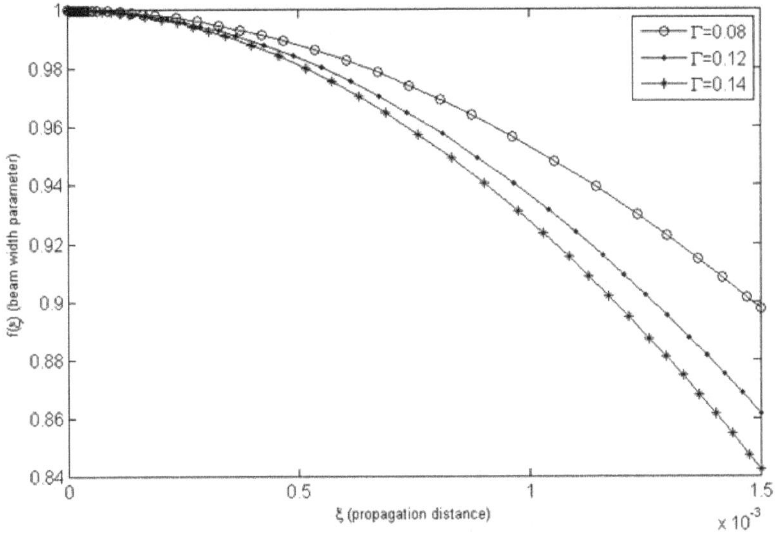

Figure 3. Variation of beam-width parameter f with propagation distance ξ for different powers of pump laser beam Γ = 0.08, 0.12, 0.14 at plasma density ω_p = 0.75 ω, b = 0.8, and r_0 = 20 μm.

Thus, a rise in laser intensity results in the propagation of a laser beam containing a greater number of relativistic electrons, leading to a higher current and, subsequently, the generation of a very strong quasi-stationary magnetic field. when a result, when the magnetic field's pinching action intensifies, the significance of the focusing effect will increase significantly. Figure 4 demonstrates the impact of altering plasma density on the relativistic self-focusing phenomenon, using a specific beginning intensity of the laser beam. Based on the data shown in Figure 4, it can be shown that the laser beam's focusing increases as the slope of the ramp density profile increases, while the corresponding focusing length reduces.

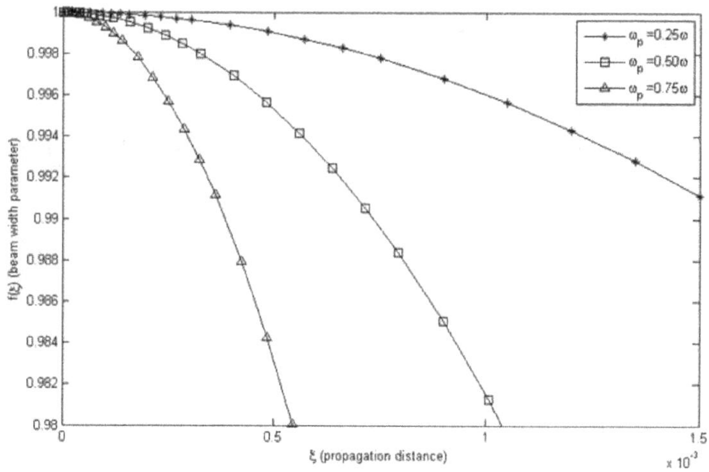

Figure 4. The effect of the change of plasma density on the relativistic self-focusing process for different densities ω_p = 0.25 ω, 0.50 ω, 0.75 ω, Γ = 0.14, b = 0.8, and r_0 = 20 μm.

It is clear that the inclusion of the quadratic r 4 term in the eikonal function modifies the radial profile. According to equation of RSF in CQP and using cost-effective decentred parameters, decreasing beam-width parameter is observed so that it could produce ultra-high laser irradiance over distances much greater than the Rayleigh length.

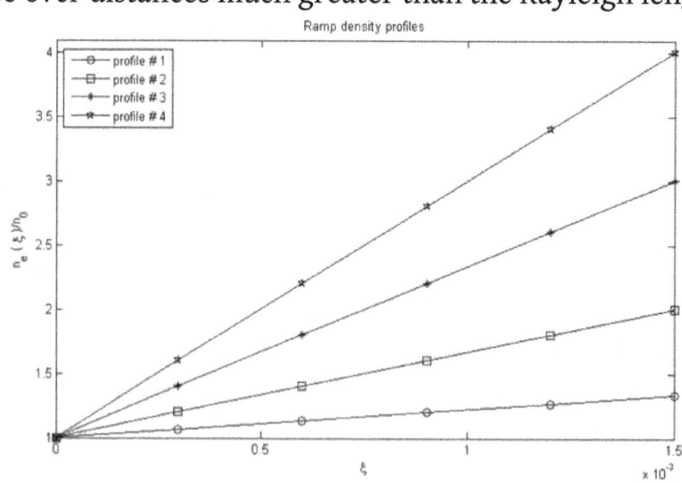

Figure 5. The normalized linear density profiles with different slopes: $n_1/n_0 \approx 1 + 6.746 \times 10^2 \tan \xi/3$ (profile#1), $n_1/n_0 \approx 1 + 33.531 \times 10^2 \tan \xi/5$ (profile#2), $n_1/n_0 \approx 1 + 40.178 \times 10^2 \tan \xi/3$ (profile#3), $n_1/n_0 \approx 1 + 20.079 \times 10^2 \tan \xi$ (profile #4).

On the other hand, we know that an upward ramp density profile as transition density gives rise more reduction in amplitude of the laser spot size close to the propagation axis [12, 25]. Therefore, we show an analysis of joint relativistic and quantum effects on ChG laser beams in one-species axial inhomogeneous cold quantum plasma (ICQP), using the high-order paraxial approach. For this purpose, Eqs. (7)–(9) should be solved simultaneously again with considering ramp density profile $\omega_p(\xi) = 0.5F(\xi)\ \omega$ and $\Gamma = 0.10, 0.12, 0.14$. Figure 5 provides information about four normalized upward density profiles with various slopes.

The linear density profile function $F(\xi) = 1 + (n_1/n_0) \tan \xi/d$ for axial inhomogeneity has been chosen. Based on this mathematical function, the slope of ramp density profile is adjustable using appropriate d and n 1 so that the range of plasma density has been adjusted from $n_c/4$ to $n_c/3$, $n_c/2$, $3n_c/4$, and n_c. The profile number of each ramp density function has been located in the top left-hand corner as a legend in Figure 5. Figures 6–9 illustrate the results of a numerical solution carried out on the self-focusing equation to assess what behaves like a high-intense ChG laser beam through inhomogeneous cold quantum plasma (ICQP).

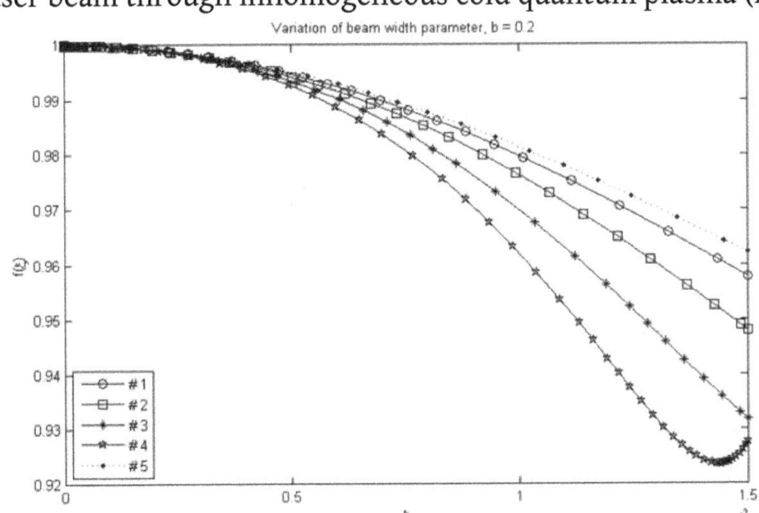

Figure 6. Variation of beam-width parameter through ICQP in the presence of ramp and uniform density profiles for b = 0.2.

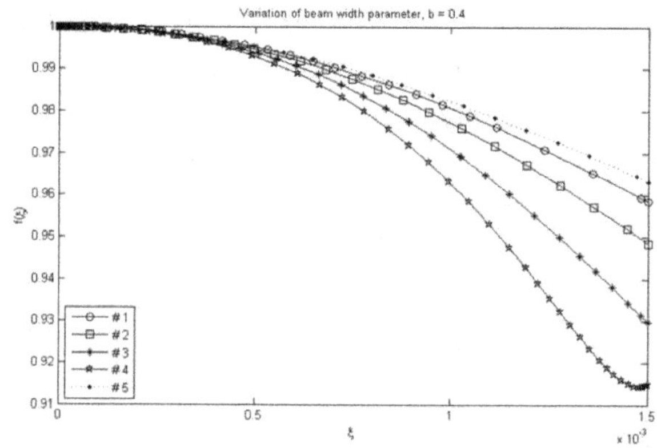

Figure 7. Variation of beam-width parameter through ICQP in the presence of ramp and uniform density profiles for b = 0.4.

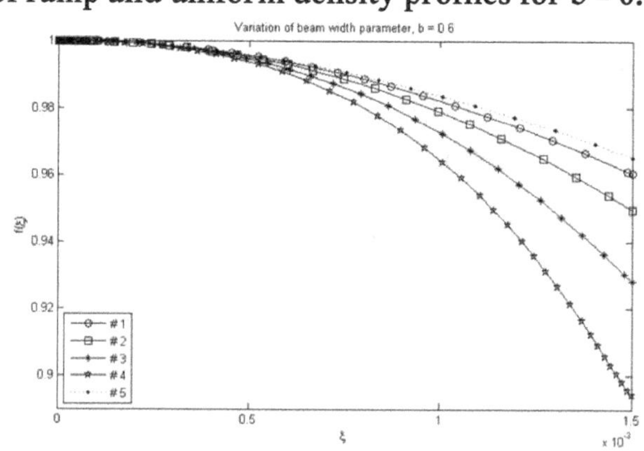

Figure 8. Variation of beam-width parameter through ICQP in the presence of ramp and uniform density profiles for b = 0.6.

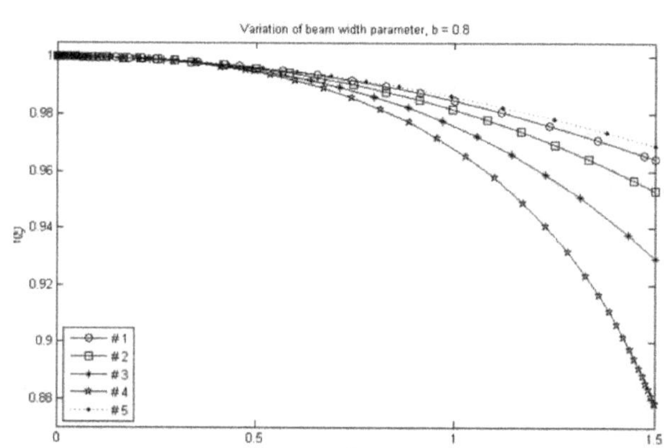

Figure 9. Variation of beam-width parameter through ICQP in the presence of ramp and uniform density profiles for b = 0.8.

In all figures, it is clear that the self-focusing becomes stronger when the slope of ramp density profile increases. Furthermore, it is noticeable that from these figures, for $\xi = 1.5 \times 10^{-3}$, the beam-width parameter decreased significantly when the higher-slope ramp density profile was used. Generally, a comparison among Figures 6–9 reveals that there was a substantial change from homogeneous to inhomogeneous quantum plasma by using various decentred parameters.

In addition, the influence of decentred parameter on the propagation of ChG laser beam in Figure 10 is illustrated for the improved density profile (profile #4). Clearly, the final values of spot size for such a laser with b from zero (Gaussian profile) to 0.8 in the $\xi = 1.5 \times 10^{-3}$ dropped significantly.

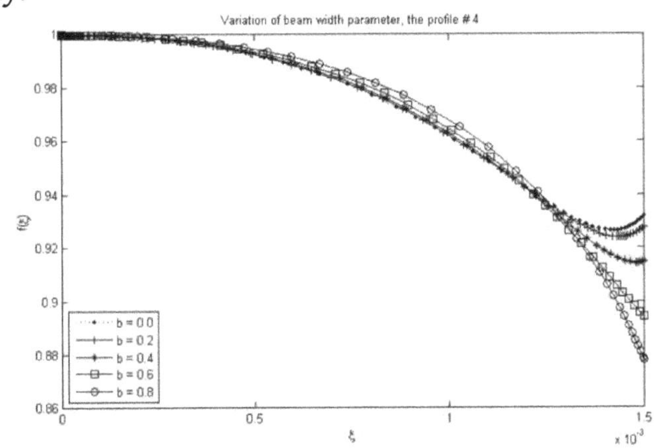

Figure 10. Variation of beam-width parameter through ICQP in the presence of ramp density profile #4 and different decentred parameters: b = 0.0 for Gaussian beam and b = 0.2, 0.4, 0.6, 0.8 for ChG beam.

Figure 11 displays various intensities to provide more detailed information and facilitate a more accurate comparison of trends when b = 0.8. The phenomenon of self-focusing in the laser beam becomes evident as the initial power of the beam increases from 0.10 to 0.14. An observable improvement in laser self-focusing is noted when a localized plasma density ramp is present in under-dense plasma. From the figure and previous ones, it is evident that the intensity of the laser, along with the type of density profile and decentred parameter, significantly contributes to improving the focusing of the ChG laser beam through the inhomogeneous cold dense plasmas.

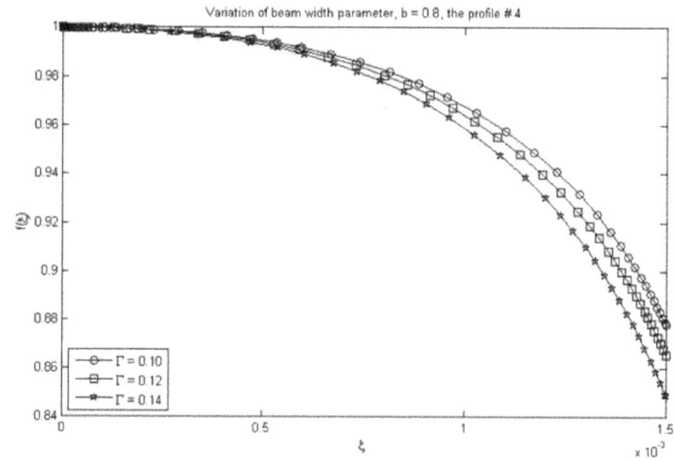

Figure 11. Variation of beam-width parameter through ICQP in the presence of ramp density profile #4, b = 0.8, and different intensities: Γ = 0.10, 0.12, 0.14.

4. CONCISE SUMMARY

Our objective was to analyze our findings on the self-focusing of high-intensity laser beams in different types of plasmas, including cold and warm quantum plasmas. Furthermore, we have implemented a ramp density profile for plasma as a significant component in enhancing self-focusing and ion acceleration. The study also examined the spread of the cosh-Gaussian intensity profile for a laser with high power and short wavelength using the higher-order paraxial approximation. The experiments reveal that the ramp density profile and off-axial contribution have a significant impact on improving the laser beam's focusing ability. The results indicate that incorporating the quadratic r^4 factor in the eikonal function alters the radial contour of the pulse. It has the potential to significantly increase the intensity of laser beams over long distances, surpassing the Rayleigh length. This can be beneficial for a wide range of laser-plasma applications.

AT A GLANCE

An important aspect in comprehending the behavior of laser beams as they propagate through plasma is the variation in spot size, which is influenced by both laser and plasma properties. Self-focusing and defocusing of the laser beam are significant factors in this investigation. This chapter aims to provide a concise examination of the prevalent phenomena in laser-plasma interaction. This paper examines the various ways in which laser beams can focus themselves through plasma. Specifically, it focuses on the self-focusing of high-intensity laser beams in quantum plasma, which is a specific type of nonlinear medium. The study provides both a brief review and a deep analysis of this phenomenon. In this scenario, the fundamental equations necessary to represent this event are constructed. Additionally, a variety of laser-plasma settings that would satisfy these equations are taken into account.

REFERENCES

1. Kaplan AE. External self-focusing of light by a nonlinear layer. Radiophysics and Quantum Electronics. 1969;12:692-696
2. Boyd RW. Nonlinear Optics. San Diego: Academic Press; 2003
3. Akhmanov SA, Khokhlov RV, Sukhorukov AP. Chapter E3 in the Laser Handbook. In: Arecchi FT, editor. E. 0. North Holland: SCHULZ-DUBOIS; 1972;2:1151
4. Boyd RW, Lukishova SG, Shen YR. Self-focusing—Past and Present. Springer; 2009. ISBN 9780387321479
5. Sprangle P, Esarey E. Physical Review Letters. Vol.67. 2021. 1991
6. Sun GZ, Ott E, Lee YC, Guzdar P. Physics of Fluids. 1987;30:526
7. Barnes D, Kurki-Suonio T, Tajima T. IEEE Transactions on Plasma Science Science. 1987;15:154
8. Kurki-Suonio T, Morrison P, Tajima T. Physical Review A. 1989;40:3230
9. Borisov A, Borovskiy A, Shiryaev V, et al. Physical Review A. 1992;45:5830
10. Mourou G, Umstadter D. Physics of Fluids. 1992;4:2315
11. Perry MD, Mourou G. Science. 1994;264:917
12. Sadighi-Bonabi R, Habibi M, Yazdani E. Physics of Plasmas. 2009;16:083105
13. Gupta DN, Hur MS, Hwang I, Suk H. Applied Physics Letters. 2007;91:081505

14. Jha P, Wadhwani N, Upadhyaya AK, Raj G. Physics of Plasmas. 2004;11:3259

15. Esarey E, Schroeder CB, Shadwick BA, Wurtele JS, Lee-mans WP. Physical Review Letters. 2000;84:3081

16. Sprangle P, Hafizi B, Penano JR. Physical Review E. 2000;61:4381

17. Hafizi B, Ting A, Hubbard RF, Sprangle P, Penano JR. Physics of Plasmas. 2003;10:1483

18. Marburger JH. Progress in Quantum Electronics. Vol. 4. Great Britain: Pergamon Press; 1975. pp. 35-110

19. Akhmanov SA, Sukhorukov AP, Khokhlov RV. Self-focusing and diffraction of light in a nonlinear medium. Soviet Physics Uspekhi. 1968;10:609

20. Sharma P. A comparison of paraxial and extended paraxial approach in laser propagation and second harmonic generation. Acta Physica Polonica. A. 2012;122:122

21. Liu CS, Tripathi VK. Laser frequency up shift, self-defocusing, and ring formation in tunnel ionizing gases and plasmas. Physics of Plasmas. 2000;7:4360

22. Liu CS, Tripathi VK. Self-focusing and frequency broadening of an intense short-pulse laser in plasmas. Journal of the Optical Society of America. A. 2001;18:1714

23. Habibi M, Ghamari F. Non-stationary self-focusing of intense laser beam in plasma using ramp density profile. Physics of Plasmas. 2011;18:103107

24. Habibi M, Ghamari F. Investigation non-stationary self-focusing of intense laser pulse in cold quantum plasma using ramp density profile. Physics of Plasmas. 2012;19:113109

25. Habibi M, Ghamari F. Stationary self-focusing of intense laser beam in cold quantum plasma using ramp density profile. Physics of Plasmas. 2012;19:103110

26. Habibi M, Ghamari F. Relativistic self-focusing of intense X-ray laser beam in warm quantum plasma with upward density profile. Physics of Plasmas. 2014;21:052705

27. Habibi M, Ghamari F. Improved focusing of a cosh-Gaussian laser beam in quantum plasma: Higher-order paraxial theory. IEEE Transactions on Plasma Science. 2015;43(7)

28. Habibi M, Ghamari F. Significant enhancement in self-focusing of high-power laser beam through dense plasmas by ramp density profile. Journal of the Optical Society of America B. 2015;32(7):1429-1434

29. Hu B, Horton W, Zhu P, Porcelli F. Physics of Plasmas. 2003;10:1015

30. Thompson MC, Rosenzweig JB, Suk H. Physical Review Accelerators and Beams. 2004;7:011301

31. Yukihiro S, Yasuhito K, Akio S, Jun A. Journal of Plasma and Fusion Research. 2004;6:715

32. Pukhov A, Kostyukov I. Physical Review E. 2008;77:025401

33. Holkundkar AR, Gupta NK. Physics of Plasmas. 2008;15:123104

34. Kim J, Kim GJ, Yoo SH. Journal of the Korean Physical Society. 2011;59:3166

35. Wang C, Li J, Sun J, Luo X. Physical Review ST-Accelerators and Beams. 2012;15. DOI: 020401

36. Chunyang Z, Zhanjun L, Shao-ping Z, Xiantu H. Journal of Plasma and Fusion Research. 2004;6:333

37. Manfredi G. How to model quantum plasmas. Fields Institute Communications. 2005;46:263

38. Shukla PK, Eliasson B. Physics-Uspekhi. 2010;53:55

39. Haas F. An introduction to quantum plasmas. Brazilian Journal of Physics. 2011;41:349

40. Shukla PK, Eliasson B. Colloquium: Nonlinear collective interactions in quantum plasmas with degenerate electron fluids. Reviews of Modern Physics. 2011;83:885

INDEX

P

Q

R

S

T

www.ingramcontent.com/pod-product-compliance
Lightning Source LLC
LaVergne TN
LVHW011850200926
848484LV00002B/431
* 9 7 8 1 7 8 7 1 5 3 8 1 3 *